Bull. Br. Mus. nat. Hist. (Geol.) **47** (1): 1–33

Neogene crabs from Brunei, Sabah and Sarawak

S. F. MORRIS

Department of Palaeontology, British Museum (Natural History), Cromwell Road, London SW7 5BD

J. S. H. COLLINS

63 Oakhurst Grove, London SE22 9AH

CONTENTS

SYNOPSIS. Thirty-six species of fossil crab are described and figured from the Neogene of Sabah, Sarawak and Brunei. The following 3 genera, 31 species and 3 subspecies are new: *Dorippe frascone tuberculata, Calappa sexaspinosa, Podophthalmus fusiformis, Charybdis feriata bruneiensis, Portunus obvallatus, P. woodwardi, Galene stipata, Parthenope (Rhinolambrus) sublitoralis, Ampliura* (gen. nov.) *simplex, Drachiella guinotae, Iphiculus granulatus, I. miriensis, I. sexspinosus, Leucosia longiangulata, L. serenei, L. tutongensis, Myra brevisulcata, M. subcarinata, M. trispinosa, Nucia borneoensis, N. calculoides, N. coxi, Nucilobus* (gen. nov.) *symmetricus, Pariphiculus gselli beetsi, P. papillosus, P. verrucosus, Philyra granulosa, Typilobus marginatus, Palaeograpsus bittneri, Pinnixa aequipunctata, P. omega, Prepaeduma* (gen. nov.) *decapoda, Xenophthalmus subitus, Macrophthalmus (Mareotis) wilfordi.*

INTRODUCTION

Most of the material forming the basis of the present descriptions was collected by the Geological Survey Departments of Brunei, Sarawak and Sabah (formerly British Territories in Borneo) during the 1950s, and sent to the British Museum (Natural History) in 1958 and 1963–64. A small quantity of material was subsequently supplied by the Brunei Shell Petroleum Company Ltd. The material is only of broad stratigraphical value because of its uniqueness in the fossil record. Reliance was placed on the published ages for the beds by foraminiferal and molluscan workers (Nuttall 1961, Haile & Wong 1965).

A general description of the geology of Sarawak, Brunei and the western part of North Borneo was published by Liechti (1960). Locality S.4918 (see p. 4), originally surveyed by the Geological Survey as Miri Formation (Pliocene) age, was later re-assessed as being of ?late middle Pleistocene age (Wilford 1961). It is probable that locality S.5545 (=S.4965) is of similar age, containing as it does *Charybdis*, an in-shore genus with a modern aspect. The Pleistocene horizons were deposited in valleys which were cut into older strata and later flooded. The Pleistocene horizons represent shallower water than the Mio-Pliocene. Barnes (1968: 337) comments: '*Macrophthalmus* is today littoral, essentially sub-tropical and tropical frequently brackish or estuarine'; there is no reason to suppose that the fossils from Brunei came from a different sort of environment.

A few leucosiids were earlier collected by members of the Geological Survey of North Borneo from isolated areas in the north-east of Borneo; they are clearly from the older Miocene deposits but their stratigraphical control is poor.

The geological ages for these specimens are given in Collenette (1954) and Haile & Wong (1965).

All the material described in this paper is deposited in the British Museum (Natural History), Department of Palaeontology: register number prefix In.

THE FAUNA AND ITS PALAEOECOLOGY

The successive Mio-Pliocene crab faunas from north-west Borneo are unusual, in that they contain an abnormally high proportion of leucosiids (species and specimens) compared with the average of the Indo-West Pacific faunas. Table 1 compares the fossil faunas with data for the Indo-West Pacific region as a whole (taken from Serène 1968: 2804 species in 462 genera) at the present day, and for the nearest comparable Recent fauna, that of the Gulf of Thailand (Rathbun 1910: 207 species in 104 genera). The Gulf of Thailand was chosen because it most closely compares with the quiet waters of Borneo and also has a higher proportion of leucosiids than the normal shelf. The Gulf of Thailand fauna was largely taken by dredge from between the coast and the off-shore fringing islands, an area protected from the high energies of the open shelf with depths ranging from shoreline to about 50 m. The bottom is largely composed of sand and broken shell. Leucosiids generally live in such areas in relatively shallow water. They are weak, sluggish back-burrowers, lurking half-buried for passing prey. It is therefore suggested that the Mio-Pliocene horizons of Borneo were deposited in similar shallow (5–50 m), low-energy areas. The lithology of the rock confirms that the bottom was of sand and broken shell, particularly suited to the back-burrowing leucosiids.

Table 1 Distribution, by percentages in family of superfamily, of crabs from the Recent of the Indo-West Pacific (n = 2804 species in 462 genera; Serène 1968), the Neogene of Borneo (n = 36 species in 22 genera; herein) and the Recent of the Gulf of Thailand (n = 204 species in 107 genera; Rathbun 1910).

	Indo-West Pacific	Borneo (Neogene)	Thailand
Dromiacea	3.4	–	3.0
Raninoidea	0.7	3.0	–
Dorippoidea	1.6	3.0	1.0
Calappidae	1.3	3.0	3.0
Leucosiidae	10.0	65.0	16.0
Majoidea	27.7	3.0	16.0
Portunoidea	7.8	9.0	10.0
Xanthoidea	22.5	3.0	22.0
Hexapodidae	4.7	3.0	13.0
Pinnotheridae	4.3	9.0	5.0
Ocypodoidea	7.3	–	6.0
Grapsoidea	11.7	3.0	10.0

In addition, there are present in Borneo more commensal species of crabs than would be expected, since many are soft-shelled and consequently easily damaged or destroyed before burial. Although there is no direct fossil evidence, annelids and holothurians probably also thrived, which would account for the presence of a higher than average proportion of species (and specimens) of crabs commensal with these animals. The tube-like enrollment of the fossil commensal crabs suggests that sideways walking was already well established.

These suggested conditions invite comparison with the crab fauna collected from the Gulf of Siam (Thailand) by the Danish Expedition to Siam (1899–1900) and described by Rathbun, 1910. She (1910: 303) notes the position of the Gulf of Siam midway between the Indian Ocean and the West Pacific Ocean, and goes on to comment on the high number of new genera and species. This suggested an area particularly suited to speciation, a comment which might be equally true for the crabs of the Pliocene of Borneo. She further noted the abundance of small forms, especially Goneplacidae (*s.l.*) and Leucosiidae, which also show the same dominance in Borneo. Initially Rathbun attributed this bias to 'the zeal of the collector' but later came to realise that it was because the collection was taken from a sheltered arm of the sea. Rathbun recorded 204 species (of which 55 are inshore/estuarine species) in her report, compared with the 36 fossil species (2 inshore/estuarine) in the Borneo fauna here reported.

Some crab groups are clearly underrepresented in the Borneo fauna, particularly the Xanthoidea and Majoidea, but in the Gulf of Siam these are either associated with a very inshore position or found in close association with algae that appear to be absent in Borneo. Serène & Soh (1976), reporting on the Brachyura collected during the Thai-Danish Expedition (1966) from approximately the same area as Rathbun's material, commented particularly on the small size of the specimens. The great majority did not exceed 10 mm. This is also true of many of the species from Borneo, so it is not considered that the Borneo faunas contain a preponderance of juvenile forms.

The two genera *Macrophthalmus* and *Charybdis* from the Pleistocene of Borneo are found in many sublittoral, estuarine or brackish waters around the Indo-Pacific, the predatory *Charybdis* being the more fully marine.

The environment during the Pliocene remained relatively steady. One species (out of 29) survived from the Lower Miri to the Seria and 10 species survived from the Lower Miri to the Upper Miri; one species that appeared in the Upper Miri survived into the Seria Formation. No crabs have been collected from the topmost series of the Borneo Pliocene, the Liang Formation.

STRATIGRAPHY

According to Van Bemmelen (1970), the geosyncline formed on the base-levelled Cretaceous was sinking by the end of the Lower Tertiary, with subsidence continuing into the Upper Tertiary. The basin was being filled by clastic sediments from the central Borneo mountainous spine to the south and from surrounding land masses to the north (Sunda Continent). Umbgrove (1933) called this late cycle geosyncline, in which it lies between stable and mobile areas resulting in weak folding, an ideogeosyncline. Haile (1969) described the formations that make up what he called the North-west Borneo Geosyncline, going on to describe its organization and evolutionary history. He compared and contrasted it with Aubouin's (1965) geosynclinal couple model. Bol & van Hoorn (1980) believed that the parallel ridges were the result of mild compressional movements related to basement wrench faulting. These weak positive areas paralleling the present north-western coastline of Borneo probably acted as barriers to strong wave action, thus having the same effect as the fringing islands along the Gulf of Thailand today.

The Miri and Seria Formations of the Belait Group underlie about 647 km² of Quaternary in Sarawak and Brunei. Outcrops are mostly to be found in sea cliffs and road cuttings. The Belait Group consists of alternations of clays, sandy clays and sandstones. The main feature-forming beds are the thick sandstone sequences, from which much of the fossil crab fauna has come. The Miri Formation is 1954 m thick at Miri and it can be lithologically and (based on smaller benthonic foraminifera) palaeontologically divided into an upper (1323 m) and a lower (631 m) sequence. The Upper Miri is the more arenaceous, with more rapid but less regular alternations. The Seria Formation (up to 2000 m thick) is conformable with the underlying Miri Formation and is structurally and lithologically similar. It can be separated palaeontologically by the appearance of the foraminifer *Triloculina* 18 (Wilford 1961: 73). The lower beds of the Seria Formation are arenaceous, giving way to sandy and silty clays in the higher beds. The Seria Formation was deposited in a very shallow seaway, with even inshore lagoonal conditions; it was certainly shallower than the preceding Miri Formation sea.

Nearly 4000 m of sediment was deposited during the late Miocene and Pliocene filling of the geosyncline. An unsubstantiated estimate by Schuppli (1946: 4) suggests that during the Neogene of north-west Borneo 50,000 ft (*c.* 15,000 m) of sediment was deposited.

The Liang Formation, which is up to 3000 ft (*c.* 920 m) thick, overlies the Seria Formation, and was deposited on an erosional surface. The Liang is predominantly a marine transgressive series of poorly consolidated sands and clays, changing upwards into similar, but lagoonal or deltaic, sediments. No fossil crabs have been collected from the Liang Formation.

The top boundary of the Liang is an unconformable erosional surface with the overlying Pleistocene terraces. The cause of this was structural, following uplift *en bloc* of the coastal region of north-west Borneo. The denudation was followed by the Jerudong Cycle, which produced a system of mature valleys in which are deposited the Jerudong Terrace sands of ?late Middle Pleistocene age, estuarine or fluviatile in character. This deposit is never more than 10 m in thickness.

LOCALITIES

J.771. Mile 3½ on Labuk Road from Sandakan, Sabah. Undifferentiated Miocene Te₅–f (?Lower Miocene) (Fitch 1958).

NB.130, NB.132. South-east part of Silimpopon region near Tawau, Sabah. Silimpopon horizon of Wenk (1938) which is a clay band lying between the Simengaris Formation and Kapilit Formation, ?Lower Miocene (Te₅–f).

NB.11541. Headwaters of Silabukan River, *c.* 14 km east of Silabukan, Sabah. Segama Group, Tungku Formation, Middle Miocene (upper Tf).

S.4807. Calcareous nodules in sea cliffs at Penanjong, 5 km north-east of Tutong, Brunei. Pliocene, Seria Formation.

S.4918. Base of marine alluvium in road cutting, Mile 3¼ on

Muara Road, 8 km north of Brunei Town, Brunei (?late middle Pleistocene – Wilford 1961: 102).

S.4965. See S.5545.

S.5536. Clay ironstone nodules in road cutting Mile 24½ on Tutong Road from Brunei Town, Brunei. Pliocene, Seria Formation.

S.5537. Clay ironstone nodules in road cutting Mile 21¼ on Tutong Road from Brunei Town, Brunei. Pliocene, Seria Formation.

S.5538. Mile 19½ on Tutong Road to Brunei Town, Brunei. Pliocene, Upper Miri Formation.

S.5539. Mile 17¾ on Tutong Road to Brunei Town, Brunei. Pliocene, Upper Miri Formation.

S.5544. Mile 13½ on Tutong Road to Brunei Town, Brunei. Pliocene, Lower Miri Formation.

S.5545 (= S.4965). Near Mile 13 on Tutong Road to Brunei Town, Brunei. Late Middle Pleistocene?

S.5548. Mile 12½ on Tutong Road to Brunei Town, Brunei. Pliocene, Lower Miri Formation.

S.5549. Mile 12¼ on Tutong Road to Brunei Town, Brunei. Pliocene, Lower Miri Formation.

S.5550. Mile 12 on Tutong Road to Brunei Town, Brunei. Pliocene, Lower Miri Formation.

S.10474. Sea cliff *c.* 1.5 km north-east of the mouth of River Trusan, south-west of Miri, Tanjong Batu area, Sarawak. Pliocene, Lower Miri Formation.

S.10475. In sea cliffs. 0.8 km north-east of the mouth of the Batang River, south-west of Miri, Sarawak. Pliocene, Lower Miri Formation but higher in succession than locality S.10474.

SYSTEMATIC PALAEONTOLOGY

Section **PODOTREMATA** Guinot, 1977

Subsection **ARCHAEOBRACHYURA** Guinot, 1977

Superfamily **RANINOIDEA** de Haan, 1841

Family **RANINIDAE** de Haan, 1841

Genus ***RANINOIDES*** H. Milne Edwards, 1837

TYPE SPECIES. By monotypy *Ranina laevis* Latreille, 1825, from the Recent (type locality unknown).

RANGE. Eocene to Recent.

Raninoides sp.   Fig. 39

MATERIAL. Two fragmentary external moulds from locality S.5539. Upper Miri Formation: In 61915 (Fig. 39), In 61916.

DESCRIPTION. The full length of neither carapace is preserved, but it probably equalled twice the maximum width which occurs at about midlength; it is moderately rounded to weakly subcarinate in transverse section and longitudinally nearly flat. Gently convex posterolateral margins converge to a posterior margin somewhat narrower than the orbitofrontal margin, which occupies about three-quarters of the width. The very short anterolateral margins terminate at the basal scars of apparently rather small, obliquely-directed spines probably weaker than the flattened outer orbital spine. Posterior to the lateral spine the lateral edges are sharply downturned and the sides are inclined almost at right angles.

There is a very small tubercle set immediately behind the lateral spine a little closer to the midline than the margin; it is seen to advantage viewed posteriorly with the carapace held at eye-level. Similarly, a pair of longitudinal nodes are seen close to the midline a little in advance of and between the anterior extremities of short, deep gastrocardiac grooves extending over the median fifth of the carapace length. The cardiac region is weakly defined between the grooves and has a transverse pair of minute tubercles at the widest part.

DISCUSSION. The greater width in proportion to the length of *Raninella toehoepae* Van Straelen 1923, from the Miocene of Borneo, distinguishes that species from *R*. sp.

Section **HETEROTREMATA** Guinot, 1977

Superfamily **DORIPPOIDEA** de Haan, 1833

Family **DORIPPIDAE** de Haan, 1833

Genus *DORIPPE* Weber, 1795

TYPE SPECIES. By subsequent designation of Latreille 1810: *Cancer quadridens* Fabricius, 1793 (= *Cancer frascone* Herbst, 1785) [ICZN Opinion 688]; from the Recent of the Indian Ocean.

RANGE. Miocene to Recent.

Dorippe (*Dorippe*) *frascone* (Herbst) *tuberculata* subsp. nov.   Fig. 1

DIAGNOSIS. Metabranchial regions more nodular than on nominal subspecies. An extra tubercle on either side of midline on the mesogastric region. Anterior part of cardiac region wider (*trans.*), and less prominent node at base of urogastric region. Sternal plates ridged with tubercles and with reduced spines on hepatic regions.

NAME. 'Tuberculate'.

HOLOTYPE. In 61853 (Figs 1a, b) from the Pliocene, Lower Miri Formation, locality S.5548. Paratypes In 61854–5 from S.5549 and In 61856 from S.5550, Lower Miri Formation.

DESCRIPTION. The carapace is subtrapezoidal in outline, almost flat longitudinally and only slightly arched transversely. The short, slightly convex anterolateral margins are barely interrupted where the cervical furrow reaches the margin; the anterior part of the posterolateral margins is nearly straight, deflected strongly outwards and, in the adult, ending in a sharp spine, but in specimens of approximately one-third the size it continues uninterrupted into the broadly rounded posterior part. A broad ridge bounds the deeply sinuous posterior margin. The orbitofrontal margin is not well preserved; it occupies rather more than half the carapace width. Basal scars indicate strong, obliquely-directed outer orbital spines and somewhat weaker spines at the lower inner orbital angle.

The cervical furrow is wide, fairly deep, and broadly V-shaped to its junction with the hepatic furrows, where it curves back to the margin; the branchiocardiac furrow, issuing from the same notch, is straighter laterally and much more rounded across the midline. Small hepatic regions are depressed and bordered above and below by low ridges,

although the anterior ridge is not developed in the younger forms. The protogastric lobes tend to coalesce medially, partially obscuring the anterior process of the small lozenge-shaped mesogastric lobe. There is a low node at the base of each protogastric lobe, a similar one on the mesogastric, one at the base of the epibranchial lobes and another occupies almost all the much reduced mesobranchial lobe. At the base of the crescentic urogastric lobe deep pits, marking the posterior gastric muscles, are bounded by low, oblique ridges. More obvious on the larger specimen is a low, isolated node at the base of the urogastric lobe; the cardiac region has a small median lobe and is more nearly flask-shaped, less rounded than on the smaller carapaces, and acutely V-shaped ridges are more prominently developed. On the meta-branchial lobes a weak groove isolates an ovate area on either side of the cardiac region and there is an obscure node anterior to the widest part of the region.

The upper surface of the larger carapace is finely pitted anteriorly, with granules grouped about the nodes; scattered granules on the branchial region become coarser and form rows parallel to the branchial furrows.

The 1st/2nd abdominal sternites are fused with the narrowly triangular, ridged 3rd sternite. The 4th sternites are subtrapezoidal in outline with much attenuated proximal angles leading down between the 5th sternites; near the anterior border a fissure extends half the distance to the midline; a steep ridge rising from the lower lateral angle recurves medially and there is a deep ovate pit on either side of the midline bordering the margin of the 5th sternite. The 5th sternite, scapuloid in outline, is bounded by a low, rounded ridge, while a more distinct median ridge forms the 'spine' terminating in a spinose 'acromion process' overlapping the 4th sternite. The rather more triangular 6th sternite has a convex basal margin and, medially, a well-rounded ridge descends steeply to the lateral margin; its 'acromion process' is formed from a secondary ridge issuing at about 45° from the median one.

DISCUSSION. This subspecies is very close to the nominal subspecies, but differs by having the tubercles along the ridges of the sternal plates. *Dorippe frascone* is distributed widely in the Indo-Pacific from East Africa to Japan and Australia. The nominal subspecies is known to live on sandy-silty or broken shell bottoms at 10–20m depth. None of the fossil specimens shows any sign of an epifauna, which is common on the Recent nominal species.

Superfamily **CALAPPOIDEA** de Haan, 1833

Family **CALAPPIDAE** de Haan, 1833

Genus *CALAPPA* Weber, 1795

TYPE SPECIES. By subsequent designation of Latreille, 1810: *Cancer granulatus* Linnaeus, 1758 [ICZN Opinion 712]; from the Recent of the Mediterranean Sea.

RANGE. Middle? Eocene to Recent.

Calappa sexaspinosa sp. nov.   Figs 2a–c

DIAGNOSIS. Carapace with six spines on the flared clypeiform posterolateral margins; the dorsal surface is tuberculate with seven of the tubercles arranged elliptically on the branchial regions.

NAME. 'Six-spined'.

HOLOTYPE. In 61857 (Figs 2a–c), a small specimen possibly a juvenile. Pliocene, Lower Miri Formation, locality S.5548. Paratypes In 61858–61 from same locality.

OTHER MATERIAL. In 61862 (S.5544), an abraded specimen only possibly assigned to this species.

DESCRIPTION. The carapace is subtrapezoidal with the beaded edge of the anterolateral margin giving way to five broadly triangular spines, increasing in size posteriorly, followed shortly by a smaller spine on the flared posterolateral margin. This flared portion is very thin and is preserved only on the left side of the type. The posterior margin is probably as wide as the front, which occupies about half the carapace width immediately in front of the flare. The broadly triangular front is downturned and constricted just above the apex; it is produced slightly beyond the outer orbital angle. The orbits are subcircular and obliquely inclined; the upper orbital margins are thickened and have two feeble notches.

Broad furrows separate the median gastric and cardiac regions from the branchial regions. On each of the protogastric lobes, which are separated from the front by a shallow transverse depression, are two tubercles, the smaller, median one slightly in advance; anterior to the tubercles is a transverse row of eight granules. There is a large tubercle with a smaller one behind it on the mesogastric lobe; one on the urogastric and two in line on the cardiac region. Of the four tubercles on each hepatic lobe, the larger anterior pair are *en échelon* with the posterior pair. The epibranchial lobe has three tubercles in a transverse line, and the mesobranchial two small ones. On the metabranchial lobes, seven tubercles are arranged more or less elliptically. Two or three granules may be scattered within the enclosed area.

Fine, pitted grooves on the metabranchial lobes extend more or less parallel with the lateral margins.

The sides are directed sharply inwards and the pleural suture and concave buccal margins are bounded by ridges.

DISCUSSION. *Calappa sexaspinosa* is closest to the Recent species *C. lophos* (Herbst, 1782) but *C. sexaspinosa* has a very different distribution of tubercles. It differs from *C. hepatica* (Linnaeus, 1758) by its greater clypeiform extension of the posterolateral margin. *C. pustulosa* Alcock, 1896 is also very similar, but this species has virtually no clypeiform extension, giving a length/breadth ratio of 1:1 as opposed to 0.7:1 for *C. sexaspinosa*.

Superfamily **PORTUNOIDEA** Rafinesque, 1815

Family **PORTUNIDAE** Rafinesque, 1815

Subfamily **PODOPHTHALMINAE** Miers, 1886

Genus *PODOPHTHALMUS* Lamarck, 1801

TYPE SPECIES. By monotypy *Podophtalmus* [sic] *spinosus* Lamarck, 1801 [= *Portunus vigil* Fabricius, 1798]; from the Recent of the Indo-Pacific Region.

RANGE. Oligocene to Recent.

Podophthalmus fusiformis sp. nov.   Figs 48–53

DIAGNOSIS. Carapace fusiform with straight, transverse upper orbital margins and well-rounded outer orbital angles; the lateral angle is at one third of the distance from the front.

NAME. 'Spindle-shaped'.

HOLOTYPE. In 62066 (Figs 48a, b). Paratypes In 62067 (Fig. 49), In 62068 (Fig. 50), In 62069 (Fig. 53), In 62070 (Fig. 51), In 62071 (Fig. 52), In 62072–96. All water-rolled internal moulds from locality S.5550, Lower Miri Formation.

DESCRIPTION. The carapace is fusiform in outline and about twice as long as broad. The rostrum (as seen on the latex mould) is very narrow, taking up only about a tenth of the orbitofrontal margin; it is steeply downturned and has a shallow median sulcus. Rather deep ocular constrictions lead to gently sinuous, almost straight upper orbital margins. The outer orbital spine was probably very weak and the lower orbital margin extends beyond the upper to the extent of the rostrum. The anterolateral margins, well rounded in front and gently concave behind, lead to strong, probably blunt spines set a little anterior to midlength, and directed slightly backwards. Sinuous posterolateral margins, rather longer than the anterolateral margins, lead by shallow coxigeal incisions to narrowly rounded posterior angles. The posterior margin is slightly convex and about half the width of the front. The cervical furrow curves forwards from immediately in front of the lateral spine, deepening at its junction with the hepatic furrow; it turns steeply backwards and inwards to the outer angle of the mesogastric lobe where it becomes deeper and more steeply inclined; at the base of the lobe it turns almost at right angles and terminates at a narrow incursion of the confluent urocardiac lobe.

The regions are distinct and slightly tumid. The mesogastric lobe is small and pentagonal; its very narrow, parallel-sided anterior process extends to the upper orbital margin and continues as an inconspicuous ridge onto the body of the lobe. A ridge on the lateral spine becomes obsolete as it progresses across the epibranchial lobe and the small, ovate mesobranchial lobes are rather more distinctly separated from the epibranchial than the metabranchial lobes. A weak branchiocardiac furrow follows the downward curve of the epibranchial ridge, reaching the margin anterior to the coxigeal incision. The urogastric and cardiac lobes together are broadly pentagonal and wider than the mesogastric lobe.

Only the male abdomen is preserved; it forms an attenuated triangle extending just into the 4th abdominal sternites; the narrow telson is about one sixth the length of the 6th somite, which is as long as the fused 4th/5th somites. The 3rd sternites are lozenge-shaped, the 4th subrectangular and the 5th–7th are chordate.

DISCUSSION. As preserved *Podophthalmus fusiformis* seems to lack the outer orbital spine of the genus. *P. vigil* (Fabricius) has orbits sloping backwards strongly, causing much reduced anterolateral margins; the lateral spines appear to be shorter than in *P. fusiformis*.

Subfamily **PORTUNINAE** Rafinesque, 1815

Genus *CHARYBDIS* de Haan, 1833

TYPE SPECIES. By subsequent designation of Glaessner, 1929: *Cancer sexdentatus* Herbst, 1783 (= *C. feriatus* Linné, 1758) [ICZN Opinion 712]; from Recent of the Indian Ocean.

RANGE. Oligocene to Recent.

Charybdis (Charybdis) feriata (Linné)
bruneiensis subsp. nov.

Figs 41, 42

1961 *Charybdis* sp. Ball, *in* Wilford: 102, 152; pl. 39 (*pars*).

DIAGNOSIS. A faint transverse ridge on the cardiac region.

NAME. 'From Brunei.'

HOLOTYPE. A part-decorticated carapace of a male, In 59015 (Figs 41a–c), from locality S.4965. Paratype In 59012 (Fig. 42), from S.4918. Both ?late Middle Pleistocene.

DESCRIPTION. The carapace is broadly ovate, the length being about four fifths of the width measured at the base of the lateral spines. In longitudinal section there is a moderate frontal depression; when viewed from the front the lateral margins are somewhat attenuated and slightly upturned. The orbitofrontal margin is about two thirds of the carapace width; no details of the front, which takes up half this distance, are preserved. Broadly ovate orbits are inclined a little outwards from the midline, and of the two notches in the slightly upturned upper orbital margin the outer is close to the outer orbital spine. Basal scars along the anterolateral margins indicate there were probably six spines. They appear to have been ovate in section, with the third about half the size of the first and succeeding pairs and the sixth thorn-like. Weakly rounded posterolateral margins lead by wide, shallow, slightly raised depressions for the 5th coxae to a broad, rounded posterior margin.

The regions are well defined, the median ones flatly tumid. On each protogastric lobe there is a short transverse ridge, almost uniting at the midline; from this ridge a triangular portion of each protogastric lobe encloses the anterior part of the semicircular mesogastric lobe. A stronger ridge crossing the broadest part of that lobe is similarly interrupted at the midline and immediately in front a low ridge develops into the anterior process which continues to the base of the small, rounded frontal lobe. Posteriorly the mesogastric lobe is divided by a furrow and its outer margins are lined with four or five granules. The urogastric lobe is represented by a small granular ridge. The cardiac region is lingulate in outline; anteriorly it is medially divided and there is a vague ridge; it becomes somewhat scabrous posteriorly. From the lateral spine a thin ridge curves the length of the epibranchial lobe. A broad depression encloses the mesobranchial lobe; between this lobe and the cardiac region is a low, rounded node and together the tumid areas form a semicircle about the mesogastric lobe.

A row of even-sized granules extends behind the upper orbital margin, there are a few atop the mesobranchial lobe and others of several diameters are scattered randomly over the dorsal surface.

On the underside deep marginal notches, giving way to shallow furrows, separate the 3rd from the 4th sternites. Shallower notches separate the 4th from the otherwise fused 5th sternites, the posterior angles of which are extremely drawn out to embrace the margin of the chordate 6th sternites; the 7th/8th are of much the same size and subreniform, while the 9th is much smaller and triangular in outline.

The right cheliped is slightly larger than the left. The anterior border of the merus has three spines of which the proximal one is the smallest. A large blunt process occurs at the inner angle of the carpus which has three spinules on the outer margin. The propodus, with five smooth costae, has two strong ridges with a spine at the distal end of each. There is also a spine on the anterior face near the carpal articulation. The fingers are about the same length as the palm.

DISCUSSION. Our species must be assigned to *Charybdis* (s.str.) because it does not have granular patches behind the epibranchial spine (*Gonioneptunus*), nor the curved posterior border of *Goniohellenus*. There are six anteriolateral teeth on the Brunei specimens, of which five are large; this rules out *Gonioinfradens* which has four large and two small. The Brunei species has only a very faint cardiac ridge; it is therefore unlikely to belong to *Goniosupradens* which has a distinct cardiac ridge. Of the 29 Recent species of *Charybdis* (*Charybdis*), 7 fall within the geographic range of *C. bruneiensis*: *C. affinis* Dana, 1852; *C. annulata* (Fabricius, 1798); *C. feriata* (Linnaeus, 1758); *C. japonica* (Milne Edwards, 1861), *C. lucifera* (Fabricius, 1798), *C. milesi* (de Haan, 1835) and *C. rosaea* (Jacquinot & Lucas, 1853). Of these species *C. japonica*, *C. annulata* and *C. lucifera* are proportionately broader than *C. bruneiensis*. The other species differ in the spinosity of the chelipeds, except for *C. feriata* which is like *C. bruneiensis* in having four spines on the merus and three on the anterior border of the propodus. Similarly *C. bruneiensis* has the characteristic notched first anterolateral spine as in *C. feriata*. They differ in that *C. bruneiensis* does not have transverse ridges on the anterior part of the mesogastric and protogastric regions, but has a faint transverse ridge on the cardiac.

Genus *PORTUNUS* Weber, 1795

TYPE SPECIES. By subsequent designation of Rathbun, 1926: *Cancer pelagicus* Linnaeus, 1758 [ICZN Opinion 394]; from Recent, type locality not known. The International Commission on Zoological Nomenclature in reaching its decision on the type species appears to have overlooked the selection by H. Milne Edwards (July 1840) of *Portunus puber* (Linnaeus, 1767).

RANGE. Miocene to Recent.

Portunus obvallatus sp. nov.

Figs 43–45

DIAGNOSIS. Carapace transversely subovate; anterolateral margins with eight spines, with the largest spine at the lateral angle; regions poorly defined and metabranchial region depressed.

NAME. 'Fortified'.

HOLOTYPE. In 61947 (♂, Fig. 43). Paratypes In 61948 (♂, Fig. 44), In 61949 (Fig 45), In 61950–6. All from locality S.5539, Upper Miri Formation. Paratype In 61957 from S.5549, Lower Miri Formation.

DESCRIPTION. A *Portunus* about half as long as broad and broadest at about midlength; moderately convex in longitudinal section, rather more steeply downturned in front and transversely almost flat. The narrowly rounded anterolateral margins have seven more or less evenly-sized granular spines followed by much larger sharp, slightly upturned spines at the lateral angles projecting straight out. The front is not well preserved; it takes up half the orbitofrontal margin, which occupies rather less than half the overall width. As far as can be made out, the outer orbital spine was weak, probably not extending beyond the front; the lower orbital margin extends beyond the thinly ridged upper orbital margin. A weak anterior-facing 'ridge' reaches only a short distance along the lateral spine and the epi- and metabranchial lobes are slightly tumid rather than ridged. The cervical furrow is straight where it crosses the midline, then turns sharply forward and becomes obsolete before reaching the margin. There is a short longitudinal groove at the base of the poorly-defined mesogastric lobe and on either side of the groove are one or two granules; the anterior mesogastric process is barely separated from the protogastric lobes. The cardiac region is broadly hexagonal and wider than the mesogastric lobe. On either side of the cardiac region is a low, rounded 'ridge' with a tubercle at its posterior end, and the metabranchial lobe is depressed against this ridge and behind the epigastric lobe.

DISCUSSION. The diminutive marginal spines anterior to the prominent one at the lateral angle distinguish *P. obvallatus* from *Portunus woodwardi* sp. nov. (below); the posterior margins of the latter are angular rather than sinuous as in *P. obvallatus* which also has a less lobate dorsal surface. The arrangement of the lateral spines is not unlike that of *Portunus sanguinolentus* (Herbst, 1769); otherwise there is no obvious comparison with other species figured from Sagami Bay by Sakai (1965) or Hawaii by Edmondson (1954).

Portunus woodwardi sp. nov.   Figs 46, 47

DIAGNOSIS. Carapace with eight anterolateral spines, the seventh vestigial and the eighth at the lateral angle; the mesogastric lobe is partially divided medially and its anterior process extends to the base of the frontal lobes.

NAME. After Dr H. Woodward, palaeontologist.

HOLOTYPE. In 61923 (♂, Figs 46a, b). Paratypes In 61924 (♂, Figs 47a, b), In 61925–36. All from Pliocene, Lower Miri Formation, of locality S.5548. Paratypes In 61937–45 from Upper Miri Formation, S.5539.

DESCRIPTION. The carapace is broader than long, with the anterolateral margins forming a broad semicircle with the front. The broadly ovate orbits take up the outer fourths of the orbitofrontal margin. The front is not well preserved on any of the available specimens. There is a single notch in the upper orbital margin and behind the short, triangular outer orbital spine are six more or less even-sized subquadrate spines followed by a smaller one which is more conspicuous on the internal mould and, for the most part, incorporated by the shell thickness into the much attenuated spine at the lateral angle. The interstices between the spines are deeply U-shaped and alternate pairs are marked by a short groove extending onto the carapace. Short posterolateral margins converge rapidly to moderate excavations for the 5th coxae and the posterior margin is weakly convex.

Low, oblique ridges separate subrectangular protogastric lobes from frontal and epigastric lobes. The hepatic region is separated by a slight furrow from the gastric region, and from the branchial region by a forwardly curved epibranchial ridge extending from the lateral spine to ovate mesobranchial lobes. The uro- and mesogastric lobes form a single, almost pentagonal area and the slender, slightly tapering anterior process reaches the base of the frontal lobes. The cardiac region is pentagonal and somewhat elongated posteriorly; anteriorly it is weakly divided medially by a furrow which extends a short way onto the urogastric lobe. The meta-branchial region has two almost confluent nodes close to the

epibranchial and cardiac borders, but is depressed laterally. There is a small ovate node tucked between the meso-branchial lobe and cardiac region.

The tumid areas of the dorsal surface are crowded with rather coarse granules which become sparser and smaller posteriorly.

On the underside of the male the 1st–3rd sternites are transversely narrowly triangular; the 4th are trapezoidal and somewhat indented by the abdominal trough; the 5th and 6th are subrectangular, tapering a little medially; the 7th are rather more rectangular, while the 8th are triangular in outline. Numerous granules crowding the surface become finer posteriorly.

One specimen, In 61924 (Figs 47a, b) has on its left side a swelling typically caused by a parasite, *Bopyrus* sp.; its remarkably large size so strongly affected the natural development of the branchiostegal areas, as well as the dorsal region, that all the anterolateral spines became completely atrophied and the lateral angle rounded. This parasite is of uncommon occurrence among fossil portunids.

DISCUSSION. While no precise details of the front of *P. woodwardi* sp. nov. are available for comparison, it is none-theless close to *Portunus arabicus* (Woodward, 1905) from the ?Pliocene of the Mekran coast, but in the latter the anterior process of the mesogastric lobe terminates at the transverse 'ridge'; the mesogastric lobe of *P. woodwardi* is proportionally larger, less rounded than that of *P. arabicus* and is divided by a median furrow. *P. woodwardi* is also close to *Portunus gladiator* Fabricius, 1798, but in the latter species anterolateral spines are more triangular and the spaces between them are V-shaped.

Superfamily **XANTHOIDEA** Dana, 1851

Family **XANTHIDAE** Dana, 1851

Genus *GALENE* de Haan, 1833

TYPE SPECIES. By monotypy *Cancer bispinosus* Herbst, 1783 [ICZN Opinion 85]; from Recent of the Indo-Pacific.

RANGE. Miocene to Recent.

Galene stipata sp. nov.   Figs 54, 55

DIAGNOSIS. *Galene* with extraorbital spine and three lateral spines.

NAME. 'Guarded'.

HOLOTYPE. In 59014 (Figs 54a–d) from the Pliocene, Lower Miri Formation of locality S.4965. Paratypes: In 61958 (Fig. 55), In 61981 from S.5548, In 61971–80 from S.5549, both Lower Miri Formation; In 61961–70 from Pliocene, Upper Miri Formation, S.5538; In 61959–60 from Pliocene, Seria Formation, S.5537.

DESCRIPTION. The anterior mesogastric process is parallel-sided for half its length; thereafter it tapers to a point terminating level with the upper orbital margins. The front follows the carapace curvature and is half the width of the orbitofrontal margin; it has a narrow median notch extending back as a groove on the dorsal surface; the inner pair of lobes are rounded, very close together and extend beyond the outer pair which form the inner orbital spines. The orbits are subovate and the wide upper orbital margin, bounded in part by a fine groove, has a beaded row of granules interrupted by two feeble notches; between the granules and groove there is a row of pits.

Where it crosses the midline, the cervical furrow is acutely V-shaped and very shallow round the base of the mesogastric lobe; from the forward angle of the mesogastric it extends in a moderately deep, broad curve forwards and outwards to the margin. The median part of the furrow becomes obsolete as growth advances and on larger specimens a false impression of its course is given by the stronger, straighter groove between the urogastric and cardiac lobes. Within the cervical furrow the posterior gastric pits are set very close together and, on specimens ranging up to at least 12.0 mm in carapace width, cuneiform pits mark the position of the internal mandible adductor muscles.

Granules of several diameters crowd the dorsal surface, those posterior slightly the larger. Sub-surface shell layers show groups of pits particularly on the more tumid areas except the cardiac region where they are arranged marginally.

The 1st–3rd abdominal somites are fused in both the male and female, the distal margin is straight, medially concave, laterally convex; a depression, broadening posteriorly, deepens from the medial concavity. A notch between the 1st–3rd and 4th sternites leads back at *c.* 65° and gives way to a broadly curved groove. The 4th sternites are rhomboidal, a broad groove runs from the distal notch to a broadly rounded posterolateral angle. The 5th sternites are half the length of the 4th and subrectangular, the 6th are somewhat longer and more quadrate, the 7th are about half the width of the 6th while the 8th are much reduced and triangular. The abdominal trough extends almost to the anterior margin of the 4th sternites. There is a scattering of granules on the 1st–3rd sternites, particularly near the margins, and on the neighbouring parts of the 4th sternites. The male abdomen tapers moderately from the 3rd somites and the telson-apex is broadly rounded. The ovate female abdomen reaches its broadest at the 4th somite; the length of the rounded-triangular telson exceeds that of the 6th somite.

DISCUSSION. Young forms are somewhat flatter in longitudinal section, more distinctly lobate and granulated and at this stage closely resemble *Lobonotus* spp. – particularly in the small node flanking the cardiac region separated from the metabranchial lobe. In *Galene stipata* this node becomes disproportionately larger as growth advances and less sharply separated from the metabranchial lobe. Also, the protogastric lobes of *G. stipata* are entire, with no tendency towards the bilobed development common to *Lobonotus*.

The present species from Brunei is very close to *Galene obscura* Milne Edwards, 1865, in that it has the extraorbital spines, a groove on the metabranchial region and a spine at the posterolateral angle. Their length/width proportions are very similar but Milne Edwards (1865) recorded five spines on the lateral margin of *G. obscura* whilst only three can be determined on *G. stipata*. *Galene stipata* differs from *Galene bispinosa* (Herbst, 1783) in all the characters stated by Milne Edwards for that species.

Superfamily **PARTHENOPOIDEA** Macleay, 1838

Family **PARTHENOPIDAE** Macleay, 1838

Genus *PARTHENOPE* Weber, 1795

TYPE SPECIES. By subsequent designation of Rathbun, 1904: *Cancer longimanus* Linnaeus, 1758 [ICZN Opinion 696]; from Recent of the Indo-Pacific Ocean.

Subgenus *RHINOLAMBRUS* Milne-Edwards, 1878

TYPE SPECIES. By original designation *Cancer contrarius* Herbst, 1804 from Recent of the East Indies.

Parthenope (Rhinolambrus) sublitoralis sp. nov. Fig. 40

DIAGNOSIS. The carapace is pentagonal, nearly as long as broad, with its margins lined with tubercles increasing in size posteriorly; the branchial regions are ridged and the dorsal surface bilaterally ornamented with unequal-sized granules.

NAME. 'Below the shore'.

HOLOTYPE. In 61917 (Figs 40a, b) from locality S.5548, Lower Miri Formation. Paratypes In 61918–22 the same locality and horizon.

DESCRIPTION. The carapace is broadly pentagonal, almost as long as broad and deeply depressed between the gastric and branchial regions and, in side view, highest at the cardiac region. There are two or three small granules on the very short anterolateral margins. The short anterior part of the posterolateral margin curves broadly with the lateral angle, and the posterior part is nearly straight; the margins are lined with seven or eight granular tubercles gradually increasing in size posteriorly. A thin ridge bounds the gently convex posterior margin and there is a small tubercle at each angle. The orbitofrontal margin is narrow and slightly elevated. As preserved the rostrum appears to have been not much produced, moderately downturned and sulcate.

From an obscure marginal notch the cervical furrow curves broadly to the angle of the mesogastric lobe, turns sharply inwards and terminates in a pit on either side of the midline. The hepatic region is small and depressed with three or four granules. Shallow constrictions separate the protogastric lobe from the small, subovate mesogastric lobe, the anterior process of which continues to the base of the rostrum. On the protogastric lobes are two rows of four granules, the innermost pair the larger; the mesogastric lobe has two large granules each flanked by a small one, and small granules flank a median one on the urogastric lobe. Several granules encircle a larger median one on the cardiac region and the oblique branchial elevations are also ornamented with unequal granules.

Parallel granulated ridges on the subhepatic and pterygostomian regions extend to the lower outer angle of the orbit. A single ridge on the branchiostegite curves in to the lower angle of the buccal margin, which is quadrate, wider than long.

Of the thoracic sternites, the first pair are very small and slightly indented from the 2nd; these are weakly delineated from the 3rd, which are subovate and deeply separated medially; together they form an obtuse triangle bounded by a raised rim. The 4th sternites are quadrate and the 5th–8th rectangular; on each of the latter is a median granule and another occurs on the outer border.

DISCUSSION. The new species is very similar to *Rhinolambrus pelagicus* (Rüppell, 1830) from the Recent of the Red Sea, but has tubercles on the branchial regions in lines separated by deep branchial grooves. Each axial region has a single large tubercle with a varying number of subsidiary tubercles, compared with the more numerous but randomly distributed tubercles on *R. pelagicus*. A Recent species from an unknown locality, *Rhinolambrus contrarius* (Herbst, 1804) has a similar distribution of tubercles but differs by having only one groove crossing the branchial region instead of the two in *R. sublitoralis*. All other species of *Rhinolambrus* have branchial processes and are therefore clearly distinct from *R. sublitoralis*. There is some similarity between the new species and *Platylambrus serratus* (H. Milne Edwards, 1834) from the Recent of the 'Indian Ocean' [*recte* west Central America, see Rathbun, 1925: 516], but the latter has the hinder part of the posterolateral margins more concave and there are a greater number of tubercles on the anterior part of the posterolateral margins.

Superfamily LEUCOSIOIDEA Samouelle, 1819

Family LEUCOSIIDAE Samouelle, 1819

Genus *AMPLIURA* nov.

TYPE SPECIES. *Ampliura simplex* gen. et sp. nov. from Pliocene, Seria Formation of Borneo.

DIAGNOSIS. Wide subcircular female abdomen, buccal cavity initially widening forwards; non-spinate, beaded lateral margins. Length to width ratio 0.85 or less. Hepatic furrows vestigial or absent.

NAME. From Latin *amplus*, large + Greek οὐρά, tail. Feminine.

DISCUSSION. Differs from *Typilobus* by its globose female abdomen and beaded lateral margins. *Nucia* also has a globose female abdomen, but differs from *Ampliura* by its spinate lateral margin and normal leucosiid buccal cavity.

In the weak development of furrows, except for the furrows bounding the cardiac region, and absence of lateral spines *Typilobus obscurus* Quayle & Collins, 1981, from the Upper Eocene of southern England, may be an early member of the genus but its abdomen is unknown; it otherwise differs by being almost circular in outline. Another species with scarcely defined regional furrows and smooth lateral margins is *Typilobus modregoi* Via Boada, 1969, but maximum width appears to occur anterior to the middle, unlike most leucosiids which have their greatest width at the middle or just posterior to it. A second species from the same horizon and locality, *Typilobus boscoi* Via Boada, 1969, resembles *Ampliura* with its female abdomen (Gómez-Alba, 1988) and vestigial hepatic grooves, but differs from it in having spinate lateral and posterior margins and normal leucosiid triangular buccal cavity.

Ampliura simplex gen. et sp. nov.   Fig. 3

DIAGNOSIS. The carapace is subovate, without lateral spines and only feebly indented at the cervical notch; the cervical and hepatic furrows are inconspicuous.

NAME. 'Simple', from the absence of any marginal processes.

HOLOTYPE. In 62157 (Fig. 3), a female from locality S.5537, Seria Formation.

DESCRIPTION. The carapace is transversely subovate with broadly rounded lateral angles, the length being about three-quarters of the breadth; transversely and longitudinally flattened and with hardly any postfrontal depression. Only the merest indentation at the cervical notch interrupts the broadly rounded anterolateral margins. The orbitofrontal margin occupies about half the carapace width; no details of the front are preserved, but the orbits are very small and circular.

The hepatic and lateral part of the cervical furrows are present as vague lines between the granules; the grooves separating the urogastric and cardiac lobes from one another and from the branchial regions, however, are more clearly defined. The mesogastric lobe is discernible rather by the grouping of granules than by any clear-cut groove. The cardiac is broader than the mesogastric lobe and more or less shield-shaped; it very slightly indents the urogastric in front and probably almost touches the posterior margin behind.

The entire dorsal surface is densely covered in flattened granules of several diameters extending over the rounded lateral edges and sides.

There is a fine groove above the subtriangular pterygostomian region; the outer shell surface near the front is missing: had it been preserved, this part of the region would in all probability have been visible in dorsal view. Oviducts open into fifth sternites.

DISCUSSION. See generic discussion.

Genus *DRACHIELLA* Guinot *in* Serène & Soh, 1976

TYPE SPECIES. By original designation *Lithadia sculpta* Haswell, 1880, from the Recent of Fitzroy I., Queensland, Australia.

Drachiella guinotae sp. nov.   Fig. 9

DIAGNOSIS. Regional grooves divide the branchial region into its component epi-, meso- and meta- regions. Cardiac region circumscribed by furrows. Protogastric region clearly differentiated. Eyes visible on dorsal surface.

NAME. In honour of Dr Danièle Guinot.

HOLOTYPE. A female carapace, In 61863 (Figs 9a–c) from *c*. 1 mile NE of R. Trusan, SW Miri, Tanjong Batu area, Sarawak; Lower Miri Formation.

DESCRIPTION. Carapace transversely subovate, length almost 0.8 of the width, widest about midlength; steeply rounded transversely and in side view steeply rounded from behind the front to the cardiac region which forms a second prominence. Distinct furrows extend from the posterolateral angles to the front, at which point they are separated by the ridged anterior mesogastric process; lobulate marginal tubercles completely isolated from all the dorsal regions. There is a deep indentation between paired tubercles at the lateral angles and one on the anterolateral margins. The straight posterolateral margins have a tubercle behind the lateral pair and another just before the ridged intestinal lobe. The intestinal lobe overhangs and gives a false impression of the posterior margin from which it is separated by a fine groove. From the anterolateral tubercle a short concave tuberculate ridge leads to the outer angle of the orbit. A thin ridge of four to five granules forms the upper orbital margin barely divided from the narrow, almost straight, bluntly depressed front. A groove leading back from the front separates prominent frontal lobes joined behind by the anterior mesogastric process. The very small circular orbits are directed obliquely upwards and divided by a strong septum from somewhat larger, deeper antennal fossae. The epistome appears to have been narrow and acutely V-shaped.

The cervical furrow is obtuse and very shallow where it

crosses the midline about mid-carapace length; becoming much broader, it turns abruptly forwards and outwards to the lateral margin; crossing the margin it divides a lineal sub-hepatic lobe and a stronger, tapering ridge on the pterygostomian region. A furrow separates obliquely inclined elliptical epigastric lobes from fairly large quadrate hepatic and triangular protogastric lobes; the latter barely separated from one another by an ill-defined furrow. A strong furrow separates the circular cardiac region from the intestinal lobe. Large epibranchial lobes are separated by a furrow (containing a row of tubercles extending from the smaller mesobranchial lobes to the cervical furrow) from a thin, rather sinuous metabranchial lobe.

Bilaterally arranged tubercles of several diameters crowd the dorsal surface; one, more conspicuous, is set behind each upper orbital margin. The more prominent of the cardiac tubercles form a saltire anteriorly with a row at its base, the whole surrounded by a ring of granules. Deep in the grooves many of the tubercles are entire – a few almost mammillate – but the vast majority are cratered with a single median granule.

The buccal margin tapers towards the front and its margin forms a line with the outer orbital margin. The pleural suture is bordered with a line of very small cratered granules while others similar to those on the dorsal surface tend to form rows on the branchiostegite. The abdominal trough is ovate and each sternite has a tubercle at its lateral edge.

DISCUSSION. The four species assigned by Guinot to the genus *Drachiella*, *D. sculpta*, *D. morum* Alcock, 1896, *D. lapillula* Alcock, 1896 and *D. aglypha* Laurie, 1906 have the branchial region entire. *D. guinotae* appears to show the primitive condition in which the subdivisions of the branchial region are defined by furrows. The protogastric, epigastric, hepatic, cardiac and intestinal regions are similarly defined by furrows. The furrow separating the hepatic from the protogastric regions is less strongly impressed than the others. The frontal region of *D. guinotae* is less wide (*trans.*) than it is in *D. morum*, and the eyes are more anteriorly situated, and therefore less visible, than in *D. morum*, but more so than in *D. lapillula*.

Genus *IPHICULUS* Adams & White, 1848

TYPE SPECIES. By monotypy *Iphiculus spongiosus* Adams & White, 1848 [ICZN Opinion 73], from Recent of the Phillipine Islands.

RANGE. Pliocene to Recent.

Iphiculus granulatus sp. nov. Figs 15, 16, 18

DIAGNOSIS. Carapace subovate with a granulated dorsal surface; no tubercles developed on the pterygostomian region.

NAME. 'Granulated'.

HOLOTYPE. In 61868 (♀, Figs 15a–c). Paratype In 61869 (Figs 16, 18). All from locality S.4807, Seria Formation.

DESCRIPTION. The outline of the carapace is closely similar to that of *Iphiculus miriensis* sp. nov. (below), but there is no overshadowing tubercle on the pterygostomian region, which is somewhat less tumid. The orbitofrontal margin is not well preserved; it occupies about two-fifths of the carapace width.

The rostrum is sulcate and there are two notches in the thin, upturned upper orbital margin.

With the exception of a narrow depression behind the front and the bases of the median furrows the dorsal surface is densely crowded with granules of several diameters.

The pleural suture is lined with granules and the underside of both sexes is similar to that of *Iphiculus miriensis* sp. nov.

Iphiculus miriensis sp. nov. Figs 11–14

DIAGNOSIS. Carapace subovate with six blunt spines on the lateral margins; the dorsal surface is ornamented with 25 small, regularly arranged tubercles.

NAME. 'From Miri'.

HOLOTYPE. In 62123 (Fig. 11). Paratypes In 62124–6 (Figs 12–14), In 62127 (♂), In 62128–30 (♀), In 62132–8 (indet. sex). All from locality S.5548, Lower Miri Formation. Paratypes In 62139–43 from S.5549, In 62144 from S.5550, all from Lower Miri Formation. Paratypes In 62121–2 from S.5539, Upper Miri Formation.

DESCRIPTION. The carapace is subovate in outline, the length being about 0.8 of the width measured between the 2nd–3rd lateral spines; longitudinally it is domed with a shallow frontal depression, and flatly domed transversely. The well-rounded anterolateral margins are armed with four blunt spines increasing in size posteriorly; the anterior one is often obscure, almost granular, and overshadowed by a large spine immediately below on the pterygostomian process. The posterolateral margins are longer than the anterolateral margins; there are two granular spines, more noticeable on young specimens, and the space between them is twice the distance which separates the foremost from the spine at the lateral angle, and the hindmost from a short, sharp spine at the posterior angle. The posterior margin is about as wide as the front, slightly concave and narrowly rimmed. The very narrow, slightly upturned front occupies about one third of the carapace width; the rostrum is small, triangular and strongly deflected downwards; the margins are upturned by a continuation of the upper orbital margins, in which there are two notches. The narrowly ovate orbits are inclined about 45° to the midline and partially separated from rather large, subcircular antennal fossae.

The cervical furrow can just be traced on some specimens; it runs slightly back for a short distance from the margin before terminating in a shallow pit, where it joins an obscure, partially developed furrow curving from the outer orbital notch and delineating the hepatic region. A broad groove separates the confluent median gastric and cardiac regions from the large, tumid intestinal region. Broad, but shallow, furrows separate the cardiac and intestinal regions from the branchial regions. There are normally 25 small tubercles on the carapace in all. There are four on each protogastric lobe, the foremost pair forming an upturned curving row with one on each hepatic region, and the hinder pair, a downward curving row with one on each epibranchial lobe; a shorter row is composed of one tubercle on each meso- and metabranchial lobe, while another two, one behind the other, on the metabranchials lie opposite two transverse cardiac tubercles; the urogastric lobe has three tubercles in an inverted triangle. Each tubercle is ringed by, and covered with, small pustular granules, while numerous granules of several diameters are scattered over the elevated parts of the carapace and extend onto the lateral spines.

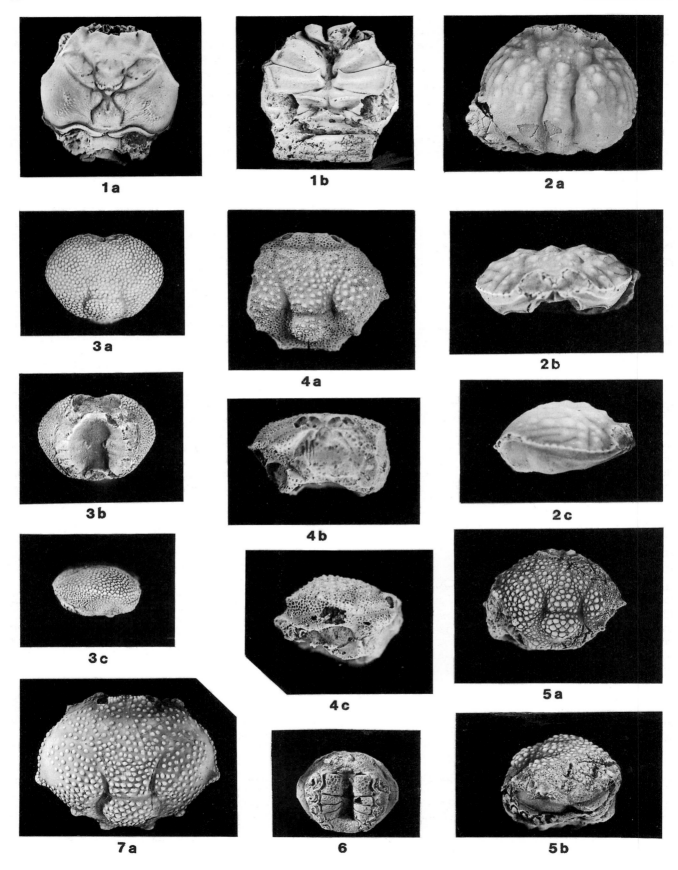

1a

1b

2a

3a

4a

2b

3b

4b

2c

3c

4c

5a

7a

6

5b

A row of granules generally lines the pleural suture, and the pterygostomian region is well delimited and tumid. The branchiostegite becomes devoid of granules posteriorly. The buccal margins are straight and divergent.

Of the sternites, the 1st and 2nd are reduced to a narrow transverse ridge with a median granular prominence; the 3rd and 4th sternites are triangular and of similar size, while the 5th–8th are subrectangular and decrease in size posteriorly.

In the male the abdominal trough extends almost the full length of the 3rd sternites and the deep, steep-sided walls are ridged above by a line of coarse granules. This ridge is not developed in the female in which the abdominal trough is broadly concave and, except anteriorly, the sternites are less studded with granules.

Iphiculus sexspinosus sp. nov.　　　　　　　　　Figs 17, 19

DIAGNOSIS. Carapace subovate with six marginal spines and a prominent pterygostomian tubercle; the cardiac and intestinal tubercles are vestigial and the dorsal tubercles are restricted to two transverse rows anteriorly.

NAME. 'Six-spined.'

HOLOTYPE. In 61864 (♂, Figs 17a, b) from locality S.5536, Seria Formation. Paratypes In 61865 (Fig. 19), In 61866 from S.5539, Upper Miri Formation; In 61867 from S.5537, Seria Formation.

DESCRIPTION. The outline of the carapace and its marginal spines is essentially similar to that of *Iphiculus miriensis* sp. nov. (p. 11). The front is slightly produced; the tip of the rostrum is obscured, but behind it is sharply divided medially by a deep V-shaped cleft which continues back a short way onto the carapace. The upper orbital margin is formed by three short equidistant spines, deeply divided by notches; the median spine is directed a little upwards.

The cardiac and intestinal tubercles are not seen on the outer shell surface, although they are vestigially present as structures on an inner-shell layer; the other dorsal tubercles are much reduced in size and tend to be restricted to two anteriorly distributed rows each comprising six tubercles.

DISCUSSION. Of the present collection *I. sexspinosus* probably comes closest to the type species *I. spongiosus*, the latter having a rather coarser surface ornament and only a single line of vague 'tubercles' corresponding to the posterior row on *I. sexspinosus*. The long lateral spine typical of *I. spongiosus* reaches a length of about a fifth of the carapace width, whilst in *I. granulatus* it would probably have reached about a quarter of the carapace width. The granulation of *I. granulatus* sp. nov. (p. 11) is coarser than that of *I. sexspinosus*, and the secondary tubercular ornament is wanting; the cardiac region is less well defined. The greater number and larger size of secondary tubercles distinguishes *I. miriensis* from *I. sexspinosus*.

Genus *LEUCOSIA* Weber, 1795

TYPE SPECIES. By subsequent designation of Holthuis, 1959: *Cancer craniolaris* Linnaeus, 1758 [ICZN Opinion 712]; from Recent of the Indo-Pacific.

RANGE. Miocene to Recent.

REMARKS. The International Commission on Zoological Nomenclature appears to have overlooked the selection by H. Milne Edwards (October, 1837: pl. 25, fig. 1) of *Leucosia urania* Fabricius, 1798 as type species.

Leucosia longiangulata sp. nov.　　　　　　　Figs 25, 26

DIAGNOSIS. The carapace is broadly rhomboidal with a narrow, slightly produced front and thin, elongate lateral angles; the deep thoracic sinus terminates in a pit from which a groove extends to the lateral margin.

NAME. From Latin *longus*, long + *angulatus*, with angles.

HOLOTYPE. In 61890 (Figs 25, 26) from locality S.10474, Lower Miri Formation.

DESCRIPTION. The carapace is rhomboidal in outline, about one-seventh longer than wide and much narrowed anteriorly. The front is slightly produced, vaguely tridentate with the sharply downturned rostrum taking up the middle third; the triangular elevation above is rounded. There is very little constriction behind the front and the gently convex antero-lateral margins lead to rather elongate lateral angles, commencing about one third distant from the front. The lateral edge is sharp and finely granulated, the granules continuing only a short distance beyond the lateral angles. The postero-lateral margins are a little recurved before acute posterior angles. The posterior margin is somewhat extended, flattened, straight and bordered with granules; it is about twice the frontal width.

The thoracic sinus is deep and ends well in front of the 1st limbs in a rather deep, obtusely ovate pit which has a narrow groove passing upward round the lateral edge immediately before the lateral angle.

DISCUSSION. See p. 15.

Leucosia serenei sp. nov.　　　　　　　　　　Figs 20–23

DIAGNOSIS. The carapace is rhomboidal with the front moderately produced and narrow; the thoracic sinus is broad and terminates in an obscure depression.

NAME. In honour of Dr Raoul Serène.

HOLOTYPE. In 61870 (Figs 20a–b). Paratypes In 61871 (Fig. 21), In 61872 (Fig. 22), In 61873 (Fig. 23), In 61874–80. All from locality S.5548, Lower Miri Formation. Paratypes In 61881–4 from S.5539, Upper Miri Formation.

Fig. 1　*Dorippe frascone tuberculata* subsp. nov. **Holotype** In 61853 from S.5548, Lower Miri Formation, × 1. a, dorsal view; b, ventral view.
Fig. 2　*Calappa sexaspinosa* sp. nov. **Holotype** In 61857 from S.5548, Lower Miri Formation, × 3. a, dorsal view; b, right lateral view; c, anterior view.
Fig. 3　*Ampliura simplex* gen. et sp. nov. **Holotype** In 62157 from S.5537, Seria Formation, × 5. a, dorsal view; b, ventral view; c, right lateral view.
Fig. 4　*Nucia borneoensis* sp. nov. **Holotype** In 62145 from S.5548, Lower Miri Formation, × 4. a, dorsal view; b, anterior view; c, right lateral view.
Figs 5, 6　*Nucia borneoensis* sp. nov. Fig. 5, paratype In 62148 from S.10475, Lower Miri Formation, × 4. a, dorsal view; b, right lateral view. Fig. 6, ventral view of paratype In 62146 from S.5548, Lower Miri Formation, × 3.
Fig. 7a　*Typilobus marginatus* sp. nov. Dorsal view of **holotype** In 62163 from NB 11541, Middle Miocene, Tungku Formation, × 3. See also Figs 7b–d (p. 14).

7c

7d

7b

9a

9b

8

9c

10

11

13

14

12

15c

15b

15a

DESCRIPTION. The carapace is rhomboidal in outline, about one-sixth longer than wide and much narrowed anteriorly. The front is moderately produced and the narrow triangular elevation above it is somewhat flattened and extends back a short distance. There is a very shallow depression either side of the front and the convex anterolateral margins are constricted again immediately before the broadly rounded lateral angles, set at some three-fifths the distance from the front. The margin is sharp and finely granulate, with the granules becoming sparser as they extend above the posterolateral margin as far as the posterior angles. The posterolateral margins converge to shallow incisions for the 5th coxae; the posterior margin is about twice the width of the front, nearly straight and bounded by a narrow, almost smooth ridge.

The thoracic sinus is very broad and ends in an obscure depression just in front of the first limb; there are a few minute granules scattered within the depression and a few others line its lower margin.

Internal moulds show low granular 'ridges' flanking an obscure median 'ridge' on the gastric region, and there is a group of three granules set in an inverted triangle on the cardiac region.

On the male abdomen somites 3–5 are fused and together equal the length of the 6th somite; the 3rd–5th somites of the female are very narrow, while the 6th occupies almost all the sternal area. In both sexes the telson is much reduced in size and extends well into the 3rd sternites.

DISCUSSION. See below.

Leucosia tutongensis sp. nov. Fig. 24

DIAGNOSIS. The carapace is rhomboidal with the front not much produced; the broad, shallow thoracic sinus terminates in a deep pit.

NAME. 'From Tutong'.

HOLOTYPE. In 61885 (Figs 24a–c) from locality S.5539, Upper Miri Formation. Paratypes In 61886 from S.5544, Lower Miri Formation and In 61887–9 from S.5548, Lower Miri Formation.

DESCRIPTION. The carapace is similar in outline to Leucosia serenei but about one-fifth longer than wide. The front is not much produced and the triangular elevation above is rounded and extends a short distance back. There is a shallow depression on either side of the front and the weakly convex anterolateral margins lead to narrowly rounded lateral angles set at about mid-carapace length. From the lateral angle a minutely granulated ridge, continuous with the anterolateral margin, extends back onto the branchial region to about as far as the 3rd pair of limbs; the true posterolateral margin is finely granulate, weakly incised for the 5th limb and leads by way of acute posterior angles to the posterior margin which is straight to weakly convex and finely granulated.

The broad, rather shallow thoracic sinus ends in a deep ovate pit just in front of the insertion of the 1st limb; a few fine granules line the lower margin of the sulcus and there is a scattering of minute granules within the pit.

Deep wedge-shaped gonopores open into the abdominal cleft at the 5th sternites and are bounded behind by a narrow, wall-like process extending from the 6th sternites (Fig. 24c). The abdomen of the male is similar to that of Leucosia serenei.

DISCUSSION of the species of Leucosia. The carapace width of Leucosia longiangulata sp. nov. is greater in relation to length than either Leucosia serenei sp. nov. or Leucosia tutongensis sp. nov.; it may be further distinguished by the elongate, forwardly situated lateral angle, by the deep thoracic sinus and the groove extending from it to the lateral edge. L. longiangulata compares well with L. vittata Stimpson, 1858, but the thoracic sinus is much further forward, hence the maximum width is also much further forward. It compares best with L. serenei sp. nov., but the thoracic sinus in the latter species does not reach the upper margin.

The subcarinate transverse section of L. serenei compares by and large with that of Leucosia obscura Bell, 1855, Recent of the Philippine Islands, but the latter species is a little broader with a narrow thoracic sinus terminating in a double notch, rather than broad and terminating in a depression as in L. serenei. L. tutongensis superficially resembles the Recent widespread Indo-Pacific Leucosia longifrons de Haan, 1841.

Genus MYRA Leach, 1817

TYPE SPECIES. By monotypy Leucosia fugax Fabricius, 1798 [ICZN Opinion 712]; from Recent of the Indo-Pacific Region.

RANGE. Miocene to Recent.

Myra brevisulcata sp. nov. Fig. 29

DIAGNOSIS. The carapace is subovate with laterally developed shallow cervical furrows; the anterolateral margin is represented by a cluster, rather than a line, of granules; larger granules are scattered over an otherwise minutely granulated surface.

NAME. 'With short furrows'

HOLOTYPE. In 61900 (Fig. 29) from locality S.10474, Lower Miri Formation.

DESCRIPTION. The carapace is subglobose with three moderately stout sharp spines, one at each posterior angle and a larger median one just above the posterior margin. The length excluding the spine slightly exceeds the breadth; it is widest at about midlength. Details of the slightly ascending front are not preserved. The outer wall of the hepatic region,

Fig. 7b–d *Typilobus marginatus* sp. nov. **Holotype** In 62163 from NB 11541, Middle Miocene, Tungku Formation, × 3. b, ventral view; c, right lateral view; d, anterior view. See also Fig. 7a (p. 12).

Fig. 8 *Typilobus* sp. Dorsal view of abraded specimen In 46373 from NB 132, ?Lower Miocene, Simengaris Formation, × 6.

Fig. 9 *Drachiella guinotae* sp. nov. **Holotype** In 61863 from S.10475, Lower Miri Formation, × 2. a, dorsal view; b, ventral view; c, anterior view.

Fig. 10 *Nucia calculoides* sp. nov. **Holotype** In 62158 from S.10474, Lower Miri Formation, × 3.

Figs 11–14 *Iphiculus miriensis* sp. nov. Locality S.5548, Lower Miri Formation. Fig. 11, dorsal view of **holotype** (♂) In 62123, × 2. Fig. 12, ventral view of paratype (♀) In 62124, × 3. Fig. 13, ventral view of paratype (♂) In 62125. Fig. 14, ventral view of paratype (♂) In 62126, × 3.

Fig. 15 *Iphiculus granulatus* sp. nov. **Holotype** (♀) In 61868 from Penanjong, Seria Formation, × 3. a–c, dorsal, ventral and anterior views.

forming the apparent anterolateral margin, is well rounded, ridged and granulated with finer granules interspersed; it terminates in a shallow pit in which the granules are minute and confined to the upper posterior wall. A line of granules overlapping the ridge passes over the pit and continues back to form the posterolateral margin, which is sharp. Its demarcating line of granules becomes finer and runs back almost as far as the middle of the cardiac region and limits the upward extent of a densely granulated area. A low, rounded ridge with scattered granules, representing the anterolateral margin, extends between the front to above the marginal pit. The cervical furrows are present as broad lateral depressions curving forward round the foregoing ridge to become obsolete just above the anterior granules of the posterolateral margins.

A few scattered granules occur on an otherwise finely pitted dorsal surface.

DISCUSSION. The extension of the posterolateral margin over the marginal pit, laterally developed cervical furrows and sparsely granulated dorsal surface, together with the broadest part of the carapace occurring further back, readily distinguish this species from *Myra subcarinata* sp. nov. and *Myra trispinosa* sp. nov.

Myra subcarinata sp. nov.   Fig. 27

DIAGNOSIS. The carapace is subovate and vaguely carinated; the median is the stoutest of the three posterior spines; the dorsal surface is finely granulated.

NAME. 'Slightly keeled'.

HOLOTYPE. In 61891 (Figs 27a–c) from locality S.5548, Lower Miri Formation. Paratype In 61892 from S.5539, Upper Miri Formation.

DESCRIPTION. The carapace is subglobose, obscurely carinated, with three stout, bluntly rounded spines, one at either angle of the posterior margin, and much the longest a median one just above the posterior margin. Exclusive of the spine, the length just exceeds the breadth. The slightly ascending orbitofrontal margin occupies about one-third of the width and the narrow triangular front is sulcate and downturned at its tip. Behind the orbital angle an angularly convex anterolateral margin is formed by the side wall of the subhepatic region which is continuous with the upper surface of the carapace and lined with a row of granules ending posteriorly in a shallow pit. Behind the pit the posterolateral margin is abruptly convex; it is sharp-edged and lined with granules for about half its length, after which the granules give way and the edge becomes rounded.

The regions are poorly defined. Behind the front, on either side of the midline is a shallow semi-circular depression. An exceedingly fine groove lined with pits separates the cardiac from the gastric regions and shallow depressions, rather than furrows, separate the anterior half of the cardiac from the branchial region.

A little above and anterior to the marginal pit is a granule somewhat larger than those crowding the dorsal surface; the latter tend to form a straggling median line extending to the front. The granules beneath the hepatic lobe and those lining either side of the marginal pit are more variable in size.

DISCUSSION. The shorter, less excavated anterolateral margin and partially granulated posterolateral margin distinguish this species from *M. fugax*.

Myra trispinosa sp. nov.   Fig. 28

DIAGNOSIS. The carapace is subovate with three stout posterior spines of which the median is the longest; the anterolateral margins are marked by a line of granules.

NAME. 'Three-spined'.

HOLOTYPE. In 61893 (Figs 28a–c) and paratypes In 61894–5, all from locality S.5548, Lower Miri Formation. Paratypes In 61896–7, from S.5539, Upper Miri Formation; paratypes In 61898–9 from S.5544, Lower Miri Formation.

DESCRIPTION. A large subglobose species with three stout, bluntly rounded spines, one at each posterior angle and a larger median one just above the posterior margin. The length, excluding spine, slightly exceeds the breadth. Details of the front are not preserved. Behind the orbital angle and above the apparent anterolateral margin formed by the outer wall of the sub-hepatic region, the true margin is represented by a short row of granules. The granules lining the hepatic margin become clustered posteriorly and the margin ends in a shallow pit. The posterolateral margin commences above the pit and its sharp, granulated edge extends almost to the posterior margin.

The regions are poorly defined. The anterior wall of the marginal pit is granulated and over the dorsal surface there is a scattering of larger granules among densely crowded minute granules. A subsurface shell layer shows a similar arrangement of pits.

DISCUSSION. *Myra trispinosa* sp. nov. is distinguishable from *M. subcarinata* by the presence of granules forming the anterolateral margin, the posterolateral margin commencing above the marginal pit, and the continued sharp edge of that margin posteriorly.

Genus *NUCIA* Dana, 1852

TYPE SPECIES. By monotypy *Nucia speciosa* Dana, 1852, from Recent of Indo-Pacific.

Figs 16, 18 *Iphiculus granulatus* sp. nov. Paratype (♂) In 61869 from Penanjong, Seria Formation, × 3. Fig. 16, dorsal view; Fig. 18, ventral view.

Fig. 17 *Iphiculus sexspinosus* sp. nov. **Holotype** In 61864 from S.5536, Seria Formation, × 2. a, dorsal view; b, ventral view.

Fig. 19 *Iphiculus sexspinosus* sp. nov. Latex cast of paratype In 61865 from S.5539, Upper Miri Formation, × 2.

Figs 20–23 *Leucosia serenei* sp. nov. from S.5548, Lower Miri Formation. Fig. 20, **holotype** In 61870, × 3. a, b dorsal and right lateral views. Fig. 21, paratype (♂) In 61871, ventral view, × 4. Fig. 22, paratype (♀) In 61872, ventral view, × 3. Fig. 23, paratype (♀) In 61873, ventral view showing the gonopores, × 3.

Fig. 24 *Leucosia tutongensis* sp. nov. **Holotype** (♂) In 61885 from S.5539, Lower Miri Formation, × 3. a, dorsal view; b, ventral view; c, right lateral view showing the gonopores.

Figs 25, 26 *Leucosia longiangulata* sp. nov. **Holotype** In 61890 from S.10474, Lower Miri Formation, × 3. Fig. 25, dorsal view; Fig. 26, right lateral view.

Fig. 27 *Myra subcarinata* sp. nov. **Holotype** In 61891 from S.5548, Lower Miri Formation, × 2. a–c, dorsal, right lateral and anterior views.

16

18

19

17a

21

22

20a

17b

23

24a

20b

25

24b

24c

27a

26

27b

27c

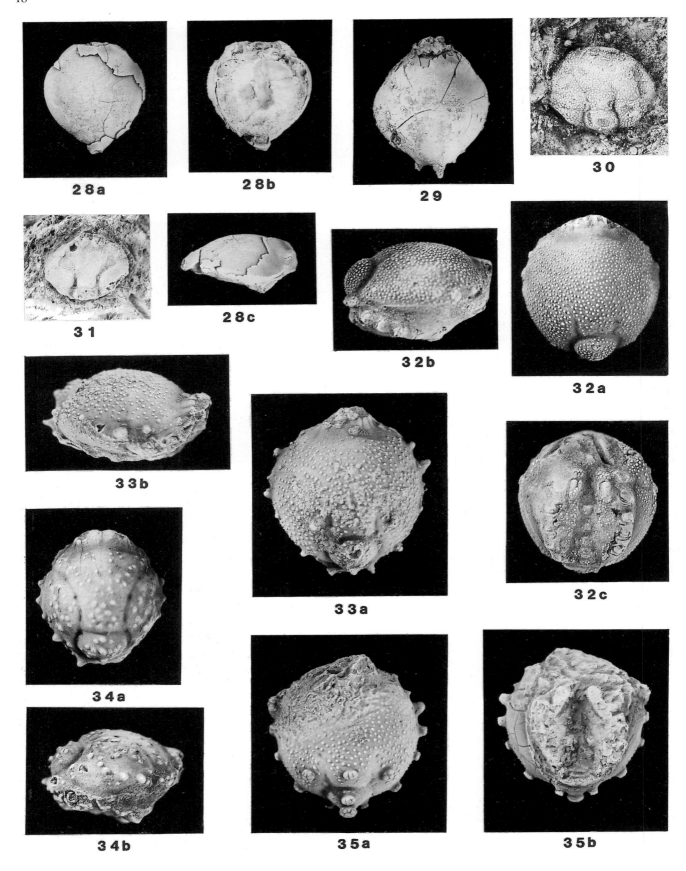

28a

28b

29

30

31

28c

32b

32a

33b

33a

32c

34a

35a

35b

34b

RANGE. Miocene to Recent.

Nucia borneoensis sp. nov.   Figs 4–6

DIAGNOSIS. The carapace is subovate with a spine at the lateral angle and one on the posterolateral margin; the furrows are fully, but shallowly, developed and the dorsal surface is coarsely granulated.

NAME. 'From Borneo.'

HOLOTYPE. In 62145 (Figs 4a–c), and paratypes In 62146 (Fig. 6), In 62147 from locality S.5548; paratypes In 62148 (Figs 5a, b), In 62149 from S.10475; paratypes In 62151–6 from S.5550. All from Lower Miri Formation.

DESCRIPTION. The carapace is transverely subovate, rather more steeply arched longitudinally than transversely. There is a sharp spine at the lateral angle, another opposite the base of the cardiac region and a small one at each corner of the intestinal region. The orbitofrontal margin occupies about half the carapace width and the circular orbits are distinctly divided by a septum from ovate antennular fossae. The broadly triangular front is steeply downturned and weakly sulcate, the sulcus extending back to a postfrontal ridge. In plan view the frontal margin is nearly straight, slightly indented at the midline.

With the exception of the well-developed furrows, which are finely and evenly granulated, the dorsal surface is densely covered with coarse granules of several diameters.

The groove above the pterygostomian process is thin and smooth, the pterygostomian region itelf is elongate, granulated and just sufficiently inflated to be level with the anterolateral margin when viewed from above.

The abdominal sternites are nearly flat on either side of the deep abdominal trough; the 4th sternites are subrectangular and about twice the length of the 5th; the 6th–8th sternites diminish in size posteriorly.

DISCUSSION. See p. 20.

Nucia calculoides sp. nov.   Fig. 10

DIAGNOSIS. The carapace is almost circular in outline, the anterolateral indentation is weak and the lateral spines are only feebly developed; the entire dorsal surface is covered by flattened granules.

NAME. Referring to the pebble-like surface ornament.

HOLOTYPE. In 62158 (Fig. 10) from locality S.10474, Lower Miri Formation. Paratypes In 62159–62 from S.5544, Lower Miri Formation.

DESCRIPTION. The carapace is almost circular in outline and moderately curved transversely. Longitudinally it is evenly curved after a shallow postfrontal depression. The anterolateral margins are convex to a shallow indentation at the cervical notch, then straight as far as a feeble spine, hardly bigger than the granules lining the edge of the margin. The posterolateral margins are slightly convex and there is no definite spine near the narrowly rounded posterior angle. The slightly convex posterior margin is about as wide as the orbitofrontal margin, which is rather less than half the carapace width; the front is a little produced, downturned and broadly sulcate. Viewed from above the frontal edge runs back on either side of a fine median notch dividing round the tip of the anterior mesogastric process. Fine granules line the frontal and orbital margins.

The grooves bounding the regions are thin, but sharply defined. The hepatic regions are clearly differentiated and the subcircular cardiac region is barely elevated above the general surface curvature.

Within the postfrontal depression there is a tendency for fine granules to form concave rows behind the orbits and affect the outline of the protogastric lobes. Behind, the dorsal surface is crowded with flattened granules interspersed with fine granules, and granules of varying size crowd the bottoms of the grooves. Similar granules crowd the elongate, rather subdued pterygostomian region as well as the groove above it and the branchiostegites.

Details of the ventral surface are not well preserved, but the abdominal trough extends to the tip of the 4th sternites, which are trapezoidal in outline and rather more coarsely granulate than the succeeding sternites.

DISCUSSION. See p. 20.

Nucia coxi sp. nov.   Figs 30, 31

1954   *Nucia* sp. Cox *in* Collenette: 15.
1954   *Nucia* sp. (probably *N. Fennemai* Böhm): Cox *in* Collenette: 15.

DIAGNOSIS. Carapace subcircular, with three lateral spines and a spine at the posterolateral angle. Cervical groove broad but shallow, becoming almost imperceptible at the midline. Dorsal surface covered with even-sized granules.

NAME. In honour of the late Dr L.R. Cox.

HOLOTYPE. In 62164 (Fig. 30) from locality J 771, (?Lower) Miocene. Paratype In 46375 (Fig. 31) from locality NB 130, ?Lower Miocene.

DESCRIPTION. The carapace is rounded in outline, somewhat broader than long; longitudinally and transversely it is gently convex. The lateral spines are not well preserved, but basal scars indicate one on the posterolateral margin behind the marginal notch, one opposite the base of the cardiac region and another equidistant between them. The spines at the posterior angles are poorly developed. The cervical furrow is broad, but shallow as it curves gently down from the marginal notch and becomes almost imperceptible where it crosses the

---

Fig. 28   *Myra trispinosa* sp. nov. **Holotype** In 61893 from S.5548, Lower Miri Formation, × 1.5. a–c, dorsal, ventral and right lateral views.
Fig. 29   *Myra brevisulcata* sp. nov. **Holotype** In 61900 from S.10474, Lower Miri Formation, × 2.
Figs 30, 31   *Nucia coxi* sp. nov. Fig. 30, **holotype** In 62164 from J.771, ?Lower Miocene, × 4. Fig. 31, paratype In 46375 from NB 130, ?Lower Miocene, × 3.
Fig. 32   *Pariphiculus gselli beetsi* subsp. nov. **Holotype** In 61901 from S.5539, Lower Miri Formation, × 4. a–c, dorsal, right lateral and ventral views.
Fig. 33   *Pariphiculus papillosus* sp. nov. **Holotype** In 61902 from S.10474, Lower Miri Formation, × 3. a, b, dorsal and right lateral views.
Fig. 34   *Nucilobus symmetricus* gen. et sp. nov. **Holotype** In 61903 from S.5549, Lower Miri Formation, × 4. a, b, dorsal and right lateral views.
Fig. 35a, b   *Pariphiculus verrucosus* sp. nov. **Holotype** In 61904 from S.10475, Lower Miri Formation, × 3. a, b, dorsal and ventral views. See also Fig. 35c (p. 22).

midline; it is joined by fine, very short, straight hepatic furrows. On each protogastric lobe an obscure ridge (more noticeable on an internal mould) runs parallel to the cervical furrow amd has a node at the posterior end; similar-sized nodes occur close to the midline on the urogastric lobe. Deep furrows separate the urogastric lobe and cardiac region from one another and from the branchial regions. The tumid, rounded-pentagonal cardiac region does not overhang the posterior margin; a narrow median tongue extends into the urogastric lobe. The dorsal surface is densely crowded with more or less even-sized granules.

DISCUSSION of species of *Nucia*. There is a noticeable difference in the length/width ratios between *N. borneoensis* and *N. calculoides*, the latter being rather wider in relation to length. It differs also in having the cardiac region wider than the base of the mesogastric lobe and ovate rather than rounded as in *N. borneoensis*; the primary granules on *N. borneoensis* are more widely dispersed and hemispherical rather than being densely packed and flattened as they are in *N. calculoides*.

The weak lateral extension of the cervical furrow and hepatic furrows, together with the finer, denser granulation, distinguishes *N. coxi* from the foregoing species. *N. baripadensis* Bachmayer & Mohanti, 1973 (Miocene, eastern India) is rather similar in outline to *N. coxi* but has a sharper indentation in the anterolateral margin; the anterior portions of the furrows are as weakly developed and the cardiac region of the Indian species appears to be more ovate; its surface ornament consists of numerous cratered pits – possibly the ornament of a sub-surface shell layer.

Of Recent species, the type species *N. speciosa* differs in having a rather more protruding front and areolated lobes which obscure the lateral extension of the cervical and hepatic furrows, but it has well-defined median gastric and cardiac lobes in common with *N. borneoensis* and *N. calculoides*. *Nucia tuberculosa* Milne Edwards, 1874 and *Nucia perlata* Sakai, 1965 have a rounded carapace with subdued marginal spines; the latter has well-defined hepatic and frontal lobes.

Nucia fennemai Böhm, 1922, from the ?Lower Miocene of Java (l/w = 0.8) is relatively broader than *N. coxi*, which has a l/w ratio of 0.9. Otherwise they are very similar; they each have three lateral spines and a posterolateral spine. *N. coxi* differs in that it has a wider front, the cervical groove is more strongly impressed and the dorsal surface is covered with densely crowded even-sized granules. The granules on the dorsal surface of *N. fennemai* are not so dense and become much more widely spaced on the branchial regions. *N. fennemai* is relatively narrower than both *N. borneoensis* (l/w = 0.75) and *N. calculoides* (l/w = 0.65). Both *N. borneoensis* and *N. calculoides* have much coarser ornament and that of *N. calculoides* is pebble-like.

Genus *NUCILOBUS* nov.

TYPE SPECIES. *Nucilobus symmetricus* sp. nov.

DIAGNOSIS. Carapace longer than broad, subovate with five small marginal tubercles. The cervical furrow is shallow medially, but has deep anterolateral extensions to the margin; it is continuous with a furrow delimiting the urogastric and cardiac regions. Hepatic furrows present. The dorsal surface has a tubercular ornamentation and there are three pairs of pits in the cervical/urocardiac furrows.

NAME. Varied from *Nucia*. Masculine.

Nucilobus symmetricus gen. et sp. nov.   Fig. 34

DIAGNOSIS. As for genus.

NAME. 'Symmetrical', from the bilateral arrangement of the surface ornament.

HOLOTYPE. A female, In 61903 (Figs 34a, b) from locality S.5549, Lower Miri Formation.

DESCRIPTION. The carapace is subovate in outline, a little longer than wide and widest at about midlength. The anterolateral margins are slightly constricted, both behind the orbits and before the cervical notch; they are short and lined with 4 or 5 granules. Behind the cervical notch a very small tubercle is followed by four evenly-spaced larger ones; the fourth, set opposite the cardiac region, is followed by another one at the posterior angle. The orbitofrontal margin occupies half the carapace width. It is not produced and is directed almost straight forward; viewed from the side the curvature is only a little less than the general longitudinal curvature of the carapace. The front is wide, steeply downturned into an acutely triangular rostrum, and a deep median sulcus bifurcates, with each branch partially encircling a 'frontal lobe'.

The cervical furrow is deep from the lateral margin to the outer angle of the mesogastric lobe, but becomes very faint where it crosses the midline; anteriorly weak furrows partly separate the hepatic regions from the protogastric lobes. Deep grooves separate the rather large subquadrate urogastric lobe from the cardiac region and the cardiac from the intestinal and branchial regions. The tumid, rounded-pentagonal cardiac region does not overhang the posterior margin and is about as wide as the intestinal region, on which the corner tubercles are somewhat more spiny than the marginal ones.

There is a pit at the junction of the cervical and hepatic furrows, one at the outer angle of the urogastric and another midway along that lobe. A row of four granules lining the anterior border of the urogastric lobe is followed by six granules encircling a median one. Granules forming an outer ring on the cardiac region enclose two larger ones and the remaining surface is sparsely ornamented by granules of several diameters bilaterally distributed in semicircular or linear patterns.

The pterygostomian region is very narrow and lined with granules extending beyond the anterolateral margin.

DISCUSSION. In having anterolaterally developed cervical furrows and hepatic furrows *Nucilobus symmetricus* has elements in common with *Nucia* – particularly the new forms *N. borneoensis* (p. 19) and *N. calculoides* (p. 19) described above. The new genus differs from *Nucia*, however, in being much longer than broad and in the presence of the furrow pits (although *Nucia speciosa* appears to have two pairs: one at the broadest part of the mesogastric lobe and one midlength of urogastric lobe). *Pariphiculus* (p. 21) has a similarly elongate carapace, but lacks the prominent grooves. This suggests a lineage from *Nucilobus* to *Pariphiculus* through a reduction in carapace grooves, especially the cervical groove which in *Pariphiculus* is seen only as an anterolateral notch.

Genus *PARIPHICULUS* Alcock, 1896

TYPE SPECIES. By subsequent designation of Rathbun, 1922: *Randallia coronata* Alcock & Anderson, 1894 [ICZN Opinion 73]; from Recent of the Bay of Bengal.

RANGE. Middle Miocene to Recent.

Pariphiculus gselli Beets *beetsi* ssp. nov.　　　Fig. 32

DIAGNOSIS. Narrow (*trans.*) intestinal region with extra tubercle at the posterolateral limit of the metabranchial region. Cardiac tubercles and pair of tubercles parallel to the branchiocardic groove obsolete.

NAME. In honour of Dr C. Beets.

HOLOTYPE. In 61901 (Figs 32a–c) from locality S.5539, Lower Miri Formation.

DESCRIPTION. The carapace is small, subglobose and subcircular in outline; the length slightly exceeds the breadth. Of the four tubercular processes on the anterolateral margin, the 1st and 2nd are hardly more than enlargements of the granules forming the surface ornament; the 4th, the largest, occurs at the widest part of the carapace and is followed immediately by another tubercle and one, much reduced, lies opposite the cardiac region. The orbitofrontal margin, occupying about a third of the carapace width, is only a little produced and barely upturned; the very small, acutely triangular rostrum is sharply downturned and a low, sinuous ridge borders the frontal margin.

The basal portion of the mesogastric region is defined by faint furrows which rapidly deepen posteriorly and delimit the urogastric and cardiac regions. The bulbous, almost circular cardiac region overhangs the posterior margin; it has two obscure, even-sized granules in line medially and is wider than the short intestinal region on which the corner tubercles are similar to the larger-sized lateral ones.

A narrow postfrontal area is only sparsely granulated, but the remainder of the dorsal surface is densely and finely granulated, the granules becoming a little coarser towards the metabranchial lobes. A row of granules, becoming sparser and finer posteriorly, borders the pleural suture. The pterygostomian region is weakly tumid and granulated; when viewed from above, a small tubercle at its centre is clearly visible beyond the lateral margin.

The abdominal sternites are shallowly depressed, more or less rectangular in outline and decrease in size posteriorly.

DISCUSSION. The new subspecies differs from the nominal subspecies (Beets 1950) from the Rembangian (Tf$_2$) of Java only by the absence of the pair of tubercles on the metabranchial region parallel to the cervical groove and the absence of the single tubercle on the urogastric region. It also differs from *Pariphiculus coronatus* (Alcock & Anderson, 1894) from the Recent of the Bay of Bengal by the absence of the same tubercles, but *P. gselli beetsi* subsp. nov. is relatively wider. *Pariphiculus rostratus* Alcock, 1896 from the Recent of the Malabar Coast differs from *P. g. beetsi* because in *P. rostratus* the cardiac tubercles have become extended into spines.

Pariphiculus papillosus sp. nov.　　　Fig. 33

DIAGNOSIS. The carapace is subovate, slightly longer than broad, with six small marginal tubercles; the larger, hindmost

tubercle on the cardiac region barely overlaps the posterior margin; the dorsal surface is densely granulated.

NAME. From the papillate nature of the marginal spines.

HOLOTYPE. In 61902 (Figs 33a, b) from locality S.10474, Lower Miri Formation.

DESCRIPTION. The carapace is subglobose and subovate in outline, only a little longer than broad with the greatest width occurring at about midlength. On the anterolateral margins are four small, nipple-like tubercles of which the first is much reduced in size; of two similar tubercles on the posterolateral margins, the first is very small and the second lies opposite the hinder, larger tubercle on the cardiac region. The posterior margin is about as wide as the frontal border. The orbitofrontal margin is directed a little upwards; the front takes up the middle third and the slightly produced, acutely triangular rostrum is downturned at its tip. The upper orbital margin is thin and deeply pierced by two notches.

Apart from the furrows delineating the basal part of the urogastric lobe and cardiac region, the lobes are undifferentiated. The bulbous, almost circular cardiac region has two tubercles, the hindmost of which barely overlaps the posterior margin. The short intestinal is as wide as the cardiac and the spine above each corner is finer and more attenuated than those on the lateral margins. There is a small tubercle on each metabranchial lobe close to the intestinal lobe.

With the exception of a narrow postfrontal area and a narrow strip above the 2nd–4th lateral tubercles, the dorsal surface is covered with densely crowded granules of more or less even size.

The lateral edges are rounded and the pterygostomian region is sparsely granulated.

DISCUSSION. This species differs from all other *Pariphiculus* by the papillate nature of the marginal spines.

Pariphiculus verrucosus sp. nov.　　　Fig. 35

DIAGNOSIS. The carapace is broadly ovate with four warty tubercles on the lateral margins; the dorsal surface is granulated and has seven tubercles, of which the hindmost one on the cardiac region clearly overhangs the posterior margin.

NAME. 'Warty', from the appearance of the tubercles.

HOLOTYPE. In 61904 (Figs 35a–c) from locality S.10475, Lower Miri Formation.

DESCRIPTION. The carapace is similar in outline to the previous species but widest at the anterior third. There are several granules decreasing in size on the short, convex anterolateral margins; a longer, warty tubercle occurs immediately after an obscure cervical notch, and is followed on the convex posterolateral margin by three other tubercles of which the second is somewhat smaller.

The orbitofrontal margin is damaged in the holotype, but appears to have been only a little upturned. The urogastric lobe and the cardiac and intestinal regions are separated from one another and from the branchial regions by deep grooves. When viewed from the side the subpentagonal cardiac, and to a lesser extent the narrow, bar-like intestinal regions, overlap the posterior margin. The bottoms of the furrows are smooth and except for the cardiac and intestinal regions, which are minutely granulated, the dorsal surface is crowded with coarse granules of several diameters. There are, in addition,

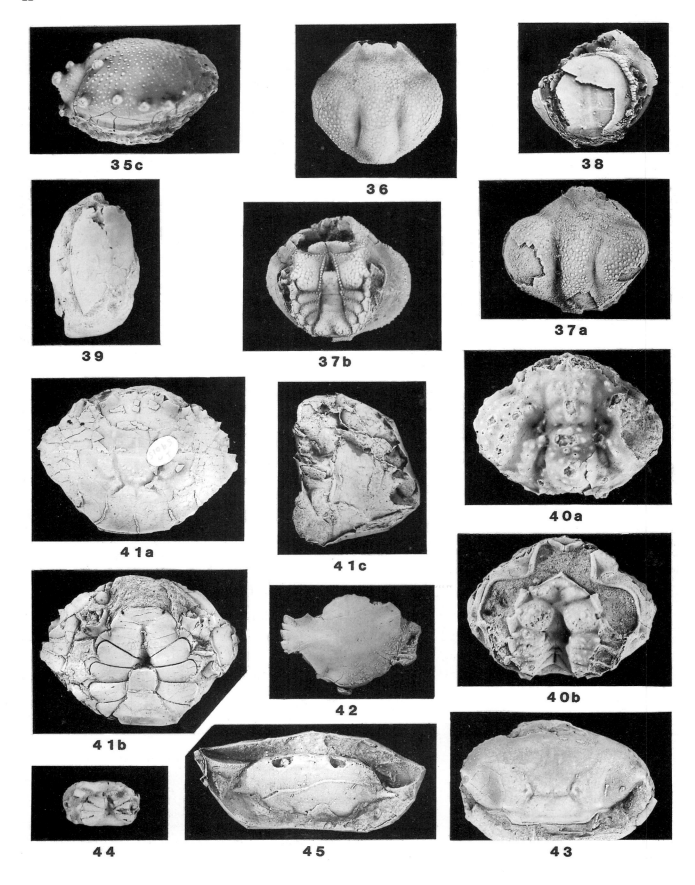

3 5 c

3 6

3 8

3 9

3 7 b

3 7 a

4 1 a

4 1 c

4 0 a

4 1 b

4 2

4 0 b

4 4

4 5

4 3

seven warty tubercles: one on the urogastric lobe; two, the foremost of which is smaller, on the cardiac region; one on each metabranchial lobe; and one at each corner of the intestinal region.

The sides are inclined a little inwards and the slightly tumid pterygostomian region has a tubercle clearly visible when viewed from above.

DISCUSSION. In the wart-like marginal and dorsal tubercles *P. verrucosus* has a marked affinity to *Pariphiculus agariciferus* Ihle, 1918, a Recent species from Roti Island (type locality) near Timor and Japan, but the latter species has additional tubercles on the gastric region, a much coarser surface granulation and poorly-defined grooves delimiting the regions.

Genus *PHILYRA* Leach, 1817

TYPE SPECIES. By subsequent designation of H. Milne Edwards, 1837: *Cancer globus* Fabricius, 1775 [ICZN Opinion 712]; from Recent of West Africa.

RANGE. Pliocene to Recent.

Philyra granulosa sp. nov.　　　　　Figs 36–38

DIAGNOSIS. Carapace octagonal with an entire lateral margin, post-frontal depression shallow; dorsal surface densely granulated.

NAME. 'Granulated'.

HOLOTYPE. In 61905 (Fig. 36), and paratypes In 61906 (Figs 37a, b), In 61907 (Fig. 38), In 61908–10 from locality S.5548, Lower Miri Formation. Paratypes In 61911–2 from S.5538, Upper Miri Formation; In 61913–4 from S.5549, Lower Miri Formation.

DESCRIPTION. The subpyriform carapace is a little longer than wide; in side view the front is a little upturned with a shallow postfrontal depression. The anterolateral margins are short, straight to slightly convex, and the beaded edge surrounding the margin is not interrupted where the lateral extension of the cervical furrow reaches the edge. The posterior margin is as wide as the orbitofrontal margin, which occupies about half the carapace width. The front is wide, triangular, steeply downturned and broadly sulcate above. The orbits are directed slightly upwards and of the two fine

fissures in the upper margins, the outermost lies close to the beaded edge.

Broad furrows separate the median gastric from the branchial regions and there is a small median incursion of the straight-sided cardiac region into the urogastric lobe; the rounded base of the cardiac region is confluent with a small, circular intestinal region. The hepatic regions are weakly tumid; deep subhepatic regions are inclined almost at right angles to the margin, and an almost smooth, flattened upper part is divided from a narrow, tumid lower part by a line of granules diverging from the marginal ones.

Numerous granules crowd the dorsal surface, although in some specimens the postfrontal depression is almost smooth; there is a denser, finer mass on either side of the cardiac region and a tendency for those on the branchial region to become coarser laterally.

In the male the 1st and 2nd sternites form a narrow triangular strip steeply inclined into the body cavity anteriorly and posteriorly sending back a narrow tongue separating the 3rd sternites. The 3rd sternites are subrectangular with a sinuous anterior border and a contra-sinuous posterior border; the subrectangular 4th sternites are wider than the 3rd and longer than the 5th–7th sternites, which become smaller and progressively more chordate posteriorly. The 4th sternites are granulated; the other sternites are relatively smooth and the abdominal trough is lined with beaded granules.

On specimen In 61906 (Figs 37a, b), a large bopyriform swelling occupies the entire left-hand branchial region.

DISCUSSION. The evenly rounded transverse section and coarsely granulated dorsal surface most readily distinguish this species; *Philyra adamsii* Bell, 1855 has granules restricted to the posterior half of the carapace, while *P. pisum* de Haan, 1841, *P. carinata* Bell, 1855 and *P. kanekoi* Sakai, 1934, all have regionally distributed granules and all have a weak median carina.

Genus *TYPILOBUS* Stoliczka, 1871

TYPE SPECIES. By monotypy *Typilobus granulosus* Stoliczka, 1871 from the Lower Miocene of Sind, India.

RANGE. Middle Eocene to Middle Miocene.

Typilobus marginatus sp. nov.　　　　　Fig. 7

DIAGNOSIS. A large, transversely ovate carapace with very thin and granulated anterolateral edge.

Fig. 35c *Pariphiculus verrucosus* sp. nov. Right lateral view of **holotype** In 61904 from S.10475, Lower Miri Formation, × 3. See also Figs 35a, b (p. 18).

Figs 36–38 *Philyra granulosa* sp. nov. From S.5548, Lower Miri Formation, × 4. Fig. 36, dorsal view of **holotype** In 61905. Fig. 37, paratype (♂) In 61906. a, b, dorsal and ventral views of male with bopyriform swelling of left branchial chamber. Fig. 38, ventral view of paratype (♀) In 61907.

Fig. 39 *Raninoides* sp. In 61915 from S.5539, Lower Miri Formation, × 2.

Fig. 40 *Parthenope (Rhinolambrus) sublitoralis* sp. nov. **Holotype** In 61917 from S.5548, Lower Miri Formation, × 3. a, b, dorsal and ventral views.

Figs 41, 42 *Charybdis (Charybdis) feriata* (Linn.) *bruneiensis* subsp. nov. from S.4965, ?late Middle Pleistocene. Fig. 41, **holotype** In 59015. a, b, dorsal and ventral views, ×1; c, latex cast from external mould to show anterolateral spines, × 2. Fig. 42, dorsal view of paratype In 59012 from S.4918, ?late Middle Pleistocene, × 2.

Figs 43–45 *Portunus obvallatus* sp. nov. from S.5539, Upper Miri Formation. Fig. 43, dorsal view of **holotype** In 61947, × 3. Fig. 44, ventral view of paratype In 61948, × 2. Fig. 45, latex cast from external mould of paratype In 61949, × 3.

NAME. Referring to the sharp lateral margin.

HOLOTYPE. In 62163 (Figs 7a–d) from locality NB 11541, Tungku Formation, Middle Miocene.

DESCRIPTION. The length of the transversely oval carapace is about four-fifths of the breadth measured immediately in front of the lateral spines; it is moderately arched transversely and longitudinally flatly arched behind a weak postfrontal depression. The anterolateral margins are convex with hardly any indentation at the cervical notch. The spine at the lateral angle is set a little behind the mid-carapace length. It is bluntly rounded and upturned, and anteriorly it tapers into a very thin edge lined with granules. At the cervical notch the ridge divides; the stronger, lower branch is interrupted by a large cluster of granules on the pterygostomian region and terminates in a small tubercle beneath the orbit, while the upper branch continues to the upper orbital margin. The posterolateral margins are nearly straight with a small, wart-like tubercle opposite the widest part of the cardiac region. A similar tubercle occurs at the posterior angle and the posterior margin is about as wide as the orbitofrontal margin. The orbitofrontal margin occupies a third of the carapace width and the sides of the barely projecting front are feebly inclined to a shallow median depression. The frontal edge is lined with two or three rows of fine granules, giving way to smaller ones lining the upper orbital margin. The small orbits are rounded, the antennar fossae obliquely ovate.

The marginal parts of the cervical and hepatic furrows are barely discernible. Weak depressions separate the urogastric from the mesobranchial lobes. The pentagonal cardiac region is tumid and separated by broad grooves from the branchial regions and bounded behind by a narrow, flattened area from the posterior margin.

With the exception of a narrow beaded strip bordering the anterolateral margins and the bottom of the furrows, the dorsal surface is densely covered with small granules of several diameters.

The pterygostomian region is subtriangular, granulate and projects beyond the anterolateral margin. There are a few small granules beneath the lateral spine.

On the ventral surface the sternites are covered with granules decreasing in size posteriorly and each pair of sternites is bordered by a row of fine granules. The 4th sternites are trapezoidal in outline and about twice as long as the 5th; the 5th and 6th pairs are subrectangular and the 7th and 8th are chordate; the outer posterior angle of the 4th–6th pairs is directed backwards. The very deep abdominal trough extends well beyond the 4th sternites and is rimmed by a row of coarse granules.

DISCUSSION. This species differs from *T. granulosus* by the sharp granulose lateral margin, the smooth band on the branchial regions parallel to the lateral margin, and the two

or three rows of granules lining the frontal edge. We agree with Dr P. Müller (*in litt.* 6/12/1987) that *Nucia baripadensis* Bachmayer & Mohanti, 1973 is probably the same as *T. granulosus*.

Typilobus sp.   Fig. 8

MATERIAL. An abraded internal mould In 46373 (Fig. 8). Locality NB 132, ?Lower Miocene Te$_5$–f, Simengaris Formation (Silimpopon horizon of Wenk, 1938). South-east part of Silimpopon area, Tawau, Cowie Harbour, Sabah.

REMARKS. In view of the characters so well preserved on the foregoing species, it would seem unwise to give a specific name to this rather poorly preserved specimen. The outline of the carapace, together with the low flattened profile and the thin lateral edge, are strongly reminiscent of *T. marginatus* sp. nov.

Section **THORACOTREMATA** Guinot, 1977

Superfamily **GRAPSOIDEA** Macleay, 1838

Family **GRAPSIDAE** Macleay, 1838

Genus *PALAEOGRAPSUS* Bittner, 1875

TYPE SPECIES. By subsequent designation of Glaessner, 1929: *Palaeograpsus inflatus* Bittner, 1875, from Upper Eocene of Vicentino, Italy.

RANGE. Middle Eocene to Pliocene.

Palaeograpsus bittneri sp. nov.   Figs 66, 67

DIAGNOSIS. A *Palaeograpsus* with the lateral margins finely granulated and a larger granule bordering the epigastric lobe; the median part of the dorsal surface is weakly depressed.

NAME. For A. Bittner.

HOLOTYPE. In 61987 (Fig. 66), and paratypes In 61988 (Fig. 67), In 61989–93 from locality S.5538, Upper Miri Formation.

DESCRIPTION. The carapace is subquadrate, almost as broad as long. Short, rounded anterolateral margins lined with several small granules terminate at a weak notch more clearly seen in side view, where a shallow furrow extends back a little before curving to the front. Behind the notch is a larger granule followed by two or three smaller ones; the marginal edge then becomes rounded posteriorly. Long, shallow depressions for the 5th coxae lead by broadly rounded angles to the posterior margin, which is concave and about as wide as the front. The slightly produced front occupies half of the orbitofrontal margin and is almost straight on either side of a

---

Figs 46, 47 *Portunus woodwardi* sp. nov. From S.5548, Lower Miri Formation, × 2. Fig. 46, **holotype** (♂) In 61923. a, b, dorsal and ventral views. Fig. 47, paratype (♂) In 61924. a, b, dorsal and ventral views showing bopyriform swelling on the left branchial chamber.

Figs 48–53 *Podophthalmus fusiformis* sp. nov. from S.5550, Lower Miri Formation. Fig. 48, **holotype** In 62066, × 2. a, b, dorsal and anterior views. Fig. 49, dorsal view of paratype In 62067, × 3. Fig. 50, ventral view of paratype In 62068, × 3. Fig. 51, dorsal view of paratype In 62070, × 3. Fig. 52, dorsal view of paratype In 62071, × 3. Fig. 53, dorsal view of paratype In 62069, × 3.

Figs 54, 55 *Galene stipata* sp. nov. Fig. 54, **holotype** In 59014 from S.4965, Lower Miri Formation, × 1. a–d, dorsal, ventral, left lateral and anterior views. Fig. 55, anteroventral view to show right cheliped of paratype In 61958 from S.5548, Lower Miri Formation, × 2.

Figs 56a, b *Prepaeduma decapoda* gen. et sp. nov. **Holotype** In 61994 from S.5549, Lower Miri Formation, × 3. a, b, dorsal and ventral views. See also Figs 56c, d (p. 27).

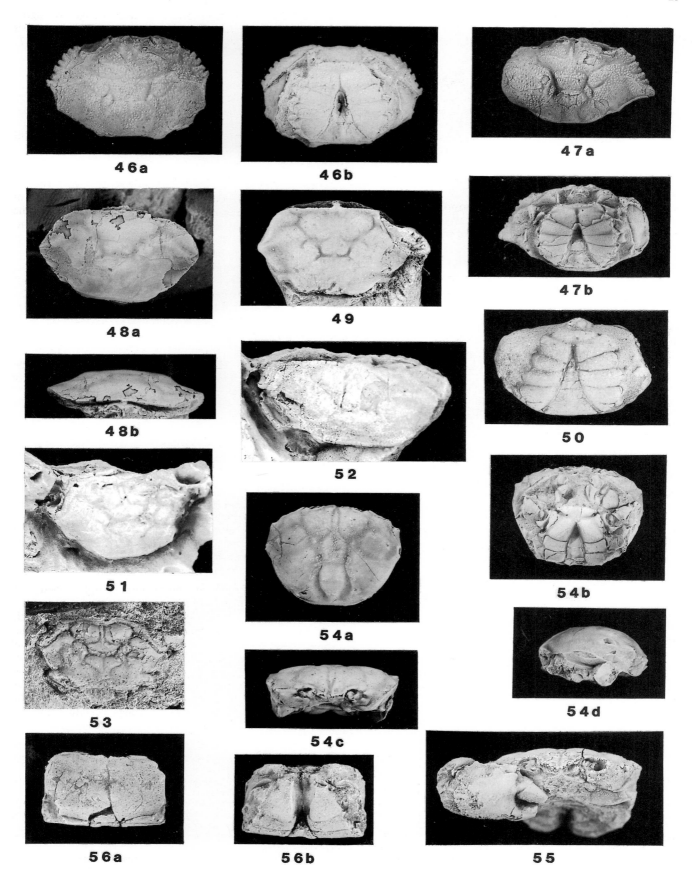

46a

46b

47a

48a

49

47b

48b

52

50

51

54a

54b

53

54c

54d

56a

56b

55

slight median notch, emphasised above by weak epigastric lobes and bordered by a fine ridge continuing round the upper orbital margin. The upper orbital margin is weakly sinuous and terminates in an inconspicuous spine.

The postcervical furrow is strongest where it crosses the midline in a gentle curve; weakening at the outer angles of the mesogastric lobe, it extends forwards and outwards and becomes obsolete before reaching the margin. Very short epimeral adductor muscle scars separate the urocardiac from the branchiocardiac regions. Only the anterior process of the mesogastric lobe is clearly defined and internal moulds show a granule at the lower inner angle of the protogastric lobes. A depression across the base of the mesogastric and basal portions of the protogastric lobes appears deepened posteriorly by tumid mesobranchial lobes and a rounded ridge on the broadly pentagonal, fused cardiac region. Wide, deep furrows extend from the broadest part of the cardiac region parallel with the coxigeal incisions; anterior to these furrows, a shallower branchiocardiac furrow crosses the carapace margin and curves toward the front.

The chelipeds are of much the same length, although the left is the heavier; the manus equals about half the carapace width. The perieopods are long and slender; the merus of the 4th is similar in size to that of the 2nd and 3rd and a little longer than the 5th.

The telson of the male abdomen reaches as far as the 3rd/4th sternite border; the 4th sternites are triangular and the even-sized 5th/8th are chordate in outline.

DISCUSSION. There is a remarkable similarity in the form of *Palaeograpsus bittneri* sp. nov. to *P. guerini* Via, 1959, from the Lutetian of Italy and Spain, to *P. depressus* Quayle & Collins, 1981 its possible derivative, and *P. bartonensis* Quayle & Collins, 1981, the last two from the Upper Eocene of the Hampshire Basin. In *P. bittneri*, however, the cervical furrow is reduced dorsally to a marginal notch; the postcervical is confined to the median part of the carapace and in these respects it is closer to the British species than to *P. guerini*. The front of *P. bittneri* is straighter and its outer angles sharper than in *P. guerini*, but probably not so far advanced as the similarly-shaped front of *P. depressus*.

Superfamily **PINNOTHEROIDEA** de Haan, 1833

Family **PINNOTHERIDAE** de Haan, 1833

Subfamily **PINNOTHERINAE** de Haan, 1833

Genus *PINNIXA* White, 1846

TYPE SPECIES. By original designation *Pinnotheres cylindricum* Say, 1818 [ICZN Opinion 85]; from Recent of Jekyll Island, Georgia, U.S.A.

RANGE. Oligocene to Recent.

Pinnixa aequipunctata sp. nov.   Figs 58–61

DIAGNOSIS. A *Pinnixa* with gastro-hepatic furrows more prominent than those dividing the branchial regions. There is a pit on either side of the midline and between them the cervical furrow is shallow: the dorsal surface of the branchial region is evenly pitted.

NAME. Referring to the pitted branchial regions.

HOLOTYPE. In 61983 (Figs 58a, b), and paratypes In 61984 (Fig. 59), In 61985 (Fig. 60), In 61986 (Fig. 61). All from locality S.5539, Upper Miri Formation.

DESCRIPTION. Carapace with length a little more than half the breadth, moderately rounded longitudinally and nearly flat in transverse section. The antero- and posterolateral margins are well rounded, but leave the lateral margins shortly subparallel. The margin edges are acute and thinly beaded. Ovate orbits, occupying the outer thirds of the orbitofrontal margin, are slightly indented medially by a projection of the upper orbital margin and a short length of the ocular peduncle, seen on the type, is contracted coincidentally. The front is not well preserved. The furrow separating the hepatic from the gastric regions commences immediately behind the outer orbital angle, and is deeper than its posterior extension which partly divides the branchial region. The cervical furrow crosses the midline in a broad curve and is shallower between small pits on either side of the midline. The protogastric lobes are feebly separated anteriorly and only a very shallow furrow separates the confluent, almost circular urogastric and cardiac lobes from the branchial regions. Pits of even size crowd the branchial regions.

DISCUSSION. So far, only a few fossil species of *Pinnixa* have been described; the earliest known, *P. eocenica* Rathbun, 1926 may be distinguished from *P. aequipunctata* in having a more rounded carapace outline, stronger branchial furrows and a narrower cardiac region. Rathbun (1932) described two Miocene species from California: of these *Pinnixa galliheri* differs from the Borneo species in having arcuate rather than straight lateral margins, while *P. montereyensis* is known by only one specimen showing its ventral surface. In the same paper Rathbun describes *Parapinnixa miocenica* (regarded by Via Boada, 1969, as *Pinnixa*) which also has angular lateral angles.

Pinnixa omega sp. nov.   Fig. 62

DIAGNOSIS. Carapace with deep grooves separating the mesogastric region from the protogastric region and the protogastric region from the hepatic lobes.

Figs 56c, d, 57 *Prepaeduma decapoda* gen. et sp. nov. Fig. 56c, d, **holotype** In 61994 from S.5549, Lower Miri Formation, × 3. c, d, left lateral and posteroventral views. See also Figs 56a, b (p. 25). Fig. 57, ventral view of paratype In 61996 from S.5550, Lower Miri Formation, × 3.

Figs 58–61 *Pinnixa aequipunctata* sp. nov. Fig. 58, **holotype** In 61983 from S.5539, Upper Miri Formation, × 3. a, b, dorsal and left lateral views of internal mould. Fig. 59, paratype In 61984 from S.5539, Upper Miri Formation, × 3. a, b, dorsal and anterior views of internal mould. Fig. 60, right oblique anterior view to show cheliped, paratype In 61985 from S.5539, Upper Miri Formation, × 5. Fig. 61, manus of right cheliped, paratype In 61986 from S.5539, Upper Miri Formation, × 10.

Fig. 62 *Pinnixa omega* sp. nov. **Holotype** In 61982 from S.5539, Upper Miri Formation, × 3. a, b, dorsal and anterior views.

Figs 63–65 *Xenophthalmus subitus* sp. nov. from S.5539, Upper Miri Formation, × 3. Fig. 63, **holotype** (♀) In 62097. a, b, dorsal and ventral views. Fig. 64, ventral view of abdomen, paratype (♂) In 62098. Fig. 65, ventral view of abdomen, paratype (♂) In 62099.

Figs 66, 67 *Palaeograpsus bittneri* sp. nov. from S.5538, Upper Miri Formation, × 3. Fig. 66, **holotype** In 61987, latex cast of external mould of dorsal. Fig. 67, paratype In 61988, latex cast of external mould of ventral.

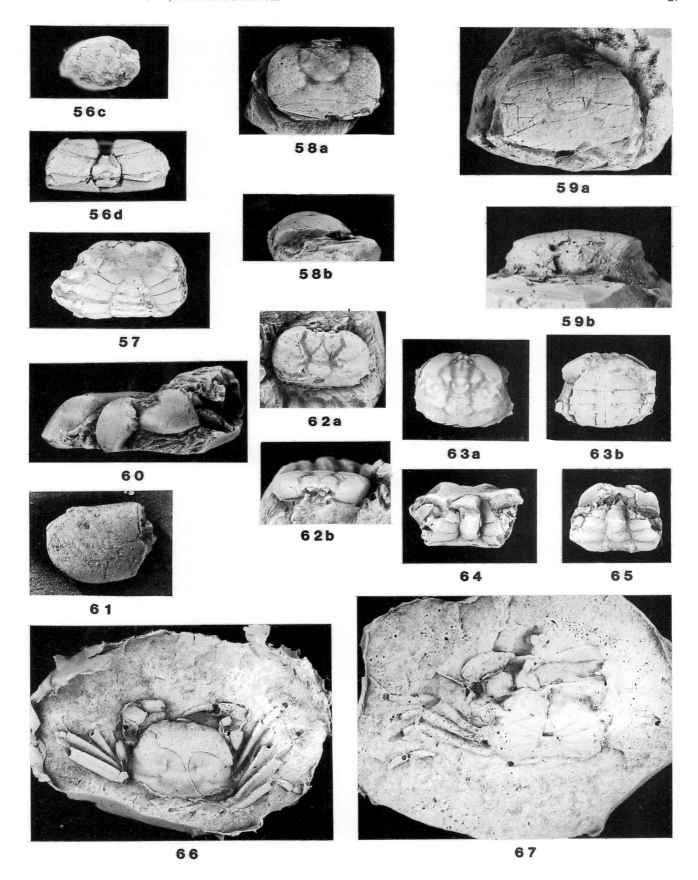

56c

56d

57

58a

58b

59a

59b

60

62a

62b

63a

63b

64

65

61

66

67

NAME. From the resemblance of the hepatic and gastric furrows to the Greek character.

HOLOTYPE. In 61982 (Figs 62a, b) from locality S.5539, Upper Miri Formation.

DESCRIPTION. The transversely subovate carapace is one-third broader than long, and nearly flat in both transverse and longitudinal sections. Small ovate orbits take up the outer fourths of the orbitofrontal margin, which occupies about a third of the carapace width. The upper orbital margin is slightly raised and the front (which is not well preserved) is emphasized above by minute frontal lobes set close to the tip of the anterior mesogastric process. The posterolateral margins are somewhat less rounded than the anterolateral margins, the posterior angle is acute and the wide posterior margin is weakly concave at its outer thirds and (possibly) convex medially; it is bounded by a fine rim. From wide, shallow marginal notches the cervical furrow extends broadly backwards and inwards. Becoming narrower, it curves sharply round the base of the hepatic region to unite with the gastro-hepatic and proto-mesogastric furrows before crossing the midline in a shallow curve; laterally it encloses a small tumid, triangular subhepatic region. From immediately behind the outer orbital angle the prominent gastro-hepatic furrow runs back for half its length, before turning angularly inwards; the proto-mesogastric furrow is only a little less wide. The hepatic regions are weakly tumid bordering the furrow, and the finely tapering anterior process of the isosceles-shaped mesogastric lobe extends to the front. The confluent urocardiac region is lingulate in outline.

DISCUSSION. The deeply incised gastric furrows and their characteristic outline immediately distinguishes P. omega from P. aequipunctata sp. nov., p. 26, and from all other species of Pinnixa. Pinnixa species are commonly commensal with annelid worms, living in their burrows and also in the burrows of worm-like holothurians.

Superfamily **HEXAPODOIDEA** Miers, 1886

Family **HEXAPODIDAE** Miers, 1886

Subfamily **HEXAPODINAE** Miers, 1886

Genus *PREPAEDUMA* nov.

TYPE SPECIES. *Prepaeduma decapoda* gen. et sp. nov. from the Pliocene of Borneo.

RANGE. Pliocene, Lower to Upper Miri Formation.

DIAGNOSIS. Hexapodid with fifth pair of pereiopods and eighth sternites fully exposed subdorsally. Female abdominal somites unfused; in the male the fourth and fifth abdominal somites are fused.

NAME. Precursor of *Paeduma*. Neuter.

DISCUSSION. Only one genus of hexapodid crab, *Paeduma* Rathbun, 1897 (= *Amorphopus* Bell, 1859 *non* Audinet-Serville, 1839) can be compared with *Prepaeduma*, because it is the only hexapodid genus bearing a fifth pereiopod, although Bell (1859) mentions in describing *Amorphopus cylindraceus* that the fifth pair of pereiopods were reduced to a mere rudiment. Bell further commented that on de Haan's figure he could detect a tubercle at the base of the fourth pereiopod which is probably a vestigial representative of the fifth pereiopod. Manning & Holthuis (1981: 174) transferred *Thaumastoplax orientalis* Rathbun, 1909, *T. chuensis* Rathbun, 1909, and an undescribed species from Japan identified by earlier workers as *T. orientalis*, from *Thaumastoplax* to *Paeduma*. Manning & Holthuis showed that all these species had the third with the fourth, and the fifth with the sixth, male abdominal segments fused, and that in all the abdomen extended forward to the posterior margin of the buccal cavity. *Thaumastoplax* is distinguished from *Paeduma* by having the third to fifth male abdominal segments fused, and the second ambulatory legs are by far the longest and strongest of the walking legs. We are unable to confirm the last character in *P. decapoda*, below, because unfortunately the legs are not preserved. *Thaumastoplax* and *Paeduma* are similar in lacking the oblique striae on the pterygostomian regions common in hexapodid genera.

As noted by Guinot (1979: 114) hexapodid crabs are generally commensal, living in the tubes of annelids and the cavities of hydrozoans. The body has become transversely elongated for ease of entry to these cavities, presumably associated with sideways walking. Guinot (1979: 115) noted that *Paeduma* seemed to conform to the twelfth [*recte*, eleventh] rule of Lankester (1904: 538–9), in which a tendency to atrophy will be seen generally in the front or rear of the tagma. In *Prepaeduma* we probably have an ancestral form in which all the segments and legs are present but in the male only abdominal somites four and five are fused.

Prepaeduma decapoda gen. et sp. nov.   Figs 56, 57

DIAGNOSIS. As genus.

NAME. 'Ten-legged'.

HOLOTYPE. In 61994 (♂, Figs 56a–d), from Pliocene, Lower Miri Formation of locality S.5549. Paratypes In 61996 (♂, Fig. 57), In 61997–99 (11 specimens) from Pliocene, Lower Miri Formation of S.5550; paratypes In 62062–5 (6 specimens) from Pliocene, Upper Miri Formation of S.5539.

DESCRIPTION. Carapace length about three-quarters of the breadth; longitudinally very convex, particularly anteriorly, and nearly flat in transverse section. The anterolateral margins are narrowly rounded and the straight posterolateral margins lead by acute posterior angles into wide incisions abutting the 7th sternites. The posterior margin is straight and about as wide as the orbitofrontal margin which takes up about two-thirds of the carapace width. The front is not well preserved; it is rounded with the longitudinal curvature of the carapace and probably did not extend beyond the outer orbital spine. Anteriorly the margin edges are acute and finely ridged; the sides are inclined more or less at right angles to the dorsal surface and the subhepatic region is slightly tumid.

The regions are poorly defined and the cervical furrow is represented by little more than lines running towards the midline from a shallow pit at the head of short, moderately deep epimeral muscle scars. The broad, subtriangular cardiac region is clearly delineated from the branchial regions. Fine granules crowd the dorsal surface.

In the male the abdominal trough is rather narrow between the 6th sternites and broadens about midlength of the 5th pair. The 5th sternites are large, subtriangular with a broadly rounded basal angle and oblique basal edge; the median edge of the 6th pair protrudes slightly beyond the 5th and its basal edge is almost straight in contrast to the indented forward edge. The 7th pair is subrectangular in outline and is the longest.

In the female a fine suture with two pits separates the 4th sternites from the 5th pair which are trapezoidal, and their rounded basal angles protrude beyond the 6th. As in the male the 7th sternites are longest and all are inclined to the midline. The subovate female abdomen is broadest across the 3rd somite and the tip of the rather large subtriangular telson, rounded apically, extends the length of the 5th sternites. Only the small rectangular 3rd and 4th/5th somites of the male abdomen are preserved; the broken anterior margin of the '5th' is about half of the broadest part of the quadrate '4th'.

The male specimen (In 61996, Fig. 57) clearly shows the 5th pereiopod and 8th sternite, both of which are situated subdorsally. An 'appareil d'accrochage du type bouton-pression' (Guinot 1979: 120) is certainly present in the male, represented by a pit in the vertical wall of the abdominal trough on the 4th sternite situated just in front of the 4th/5th sternite boundary.

DISCUSSION. *P. decapoda* is most similar to *Paeduma orientale* (Rathbun, 1909) but differs in the segments of the male abdomen. It is also comparable to *Hexapinus latipes* (de Haan, 1835), which has the third male abdominal segment fused to the fourth and fifth segments.

Subfamily XENOPHTHALMINAE Alcock, 1900

Genus *XENOPHTHALMUS* White, 1846

TYPE SPECIES. By monotypy *Xenophthalmus pinnotheroides* White, 1846, from the Recent of the Philippines.

RANGE. Pliocene to Recent.

Xenophthalmus subitus sp. nov. Figs 63–65

DIAGNOSIS. Hepatic regions project anteriorly beyond the frontal region. Front with longitudinal groove which continues between the epigastric lobes. Discontinuous transverse ridge crosses the carapace at the level of the cardiac region.

NAME. 'Sudden' or 'unexpected'.

HOLOTYPE. In 62097 (Figs 63a, b) and paratypes In 62098 (Fig. 64), In 62099 (Fig. 65), In 62100–19 from locality S.5539, Upper Miri Formation.

DESCRIPTION. The carapace is subovate, about one-fourth broader than long; moderately curved longitudinally, but downturned rather more steeply in front and nearly flat in transverse section. Short anterolateral margins are rounded continuously into the front and thinly ridged; the ridge is a little more accentuated at the outer orbital angles, but becomes obsolete behind the orbits. Anteriorly the sides are inclined at about right angles with a low ridge just above the pleural suture; they become rounded and splayed out posteriorly. Broad, rounded posterior angles lead by shallow incisions for the fifth coxae into a narrow posterior margin which is rather more steeply concave in the male. The very small orbits are obliquely ovate, in line with the longitudinal curvature of the carapace. The front is not produced, sulcate above with small terminal nodes followed by tumid, somewhat elongated epigastric lobes. A transverse, sinuous row of tubercles is formed by one median on the mesogastric lobe,

two on each protogastric and one on each epibranchial lobe; behind these, smaller granules on the mesobranchial and one on either side of the median mesogastric form a second, almost parallel row. Another granule occurs just above the coxigeal incision. Curving across the metabranchial lobes a sharp ridge is interrrupted by grooves delimiting the broadly pentagonal cardiac region and continues across its broadest part.

The broadly ovate female abdomen covers the entire ventral surface. The 6th somite is fractionally larger than the 3rd–5th somites, and at the junction of each somite there is a pit in the trough on either side of the raised median portion; on the 3rd somite is a low transverse ridge. The male abdomen is about a third the width of that of the female and parallel-sided.

DISCUSSION. The backward direction of the orbit suggests this species should be assigned to *Xenophthalmus*, but other characters on the carapace, especially the discontinuous transverse ridge level with the cardiac region, invites comparison with *Neoxenophthalmus obscurus* (Henderson, 1893) in which the ridge separates a punctate area from the smooth posterior region. The holotype (In 62097) has the peduncles of the eyes deformed, giving a misleading impression that the orbits are inclined at an oblique angle to the midline (Fig. 63a). The Brunei species has all the regions more strongly expressed than in *Xenophthalmus pinnotheroides* White, 1846, and lacks the ridge from the posterior margin crossing the metabranchial and joining the cardiac transverse ridge.

This group of hexapodids is probably commensal with annelids in relatively shallow water in the 5–30m range and on a muddy bottom.

Superfamily OCYPODOIDEA Rafinesque, 1815

Family OCYPODIDAE Rafinesque, 1815

Subfamily MACROPHTHALMINAE Dana, 1852

Genus *MACROPHTHALMUS* Latreille, *in* Desmarest 1823

TYPE SPECIES. By subsequent designation of H. Milne Edwards, May, 1841: *Gonoplax transversus* Latreille, 1817; from Recent of the Indian Ocean.

RANGE. Miocene to Recent.

Subgenus *MAREOTIS* Barnes, 1967

TYPE SPECIES. By original designation *Macrophthalmus japonicus* de Haan, 1835; from Recent of the Indo-Pacific region.

RANGE. Pliocene to Recent.

Macrophthalmus (Mareotis) wilfordi sp. nov. Figs 68–72

1961 *Macrophthalmus latreilli* (Desmarest): Wilford: 102; pl. 39.

DIAGNOSIS. The carapace is widest between the tips of the outer orbital spines; without transverse lines of granules on the dorsal surface; the fixed finger of the cheliped is depressed.

NAME. For G.E. Wilford.

HOLOTYPE. In 59000 (Figs 68a, b). Paratypes In 59001–4,

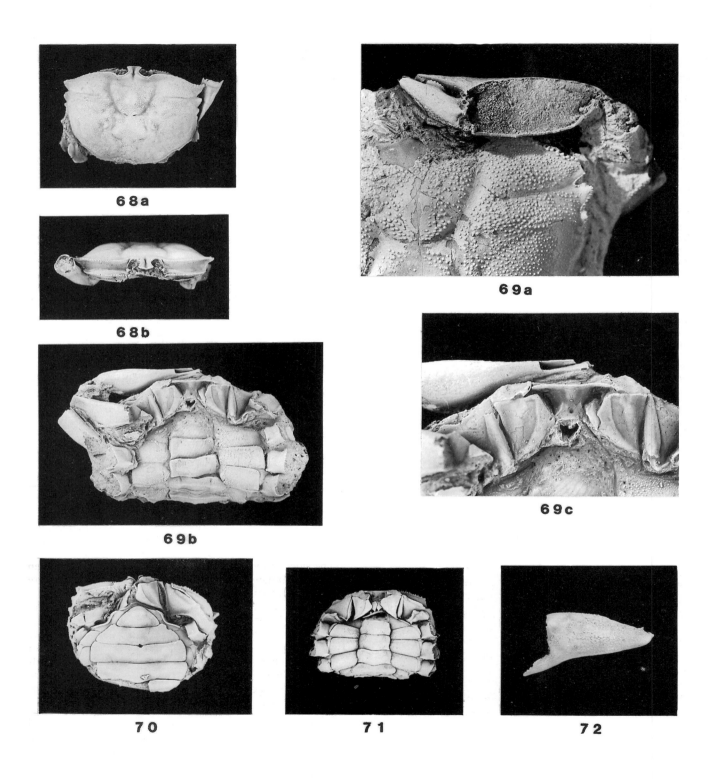

Figs 68–72 *Macrophthalmus (Mareotis) wilfordi* sp. nov. from S.4918, ?late Middle Pleistocene. Fig. 68, **holotype** In 59000, × 1. a, b, dorsal and anterior views. Fig. 69, paratype In 62120. a, anterolateral angle and incomplete cheliped, × 3. b, ventral view, × 1.5. c, epistome, × 3. Fig. 70, ventral view of abdomen, paratype (♀) In 59005, × 1.5. Fig. 71, ventral view of abdomen, paratype (♂) In 59008, × 1. Fig. 72, propodus, paratype In 61810, × 1.

Table 2 Stratigraphical distribution of fossil crab species from the Neogene of Borneo.

	Lower Miocene	Middle Miocene	Upper Miocene	Lower Miri Fm	Upper Miri Fm	Seria Fm	Middle Pleistocene
Raninoides sp.	–	–	–	–	×	–	–
Dorippe frascone (Herbst) *tuberculata* subsp. nov.	–	–	–	×	–	–	–
Calappa sexaspinosa sp. nov.	–	–	–	×	–	–	–
Podophthalmus fusiformis sp. nov.	–	–	–	×	–	–	–
Charybdis feriata (Linn.) *bruneiensis* subsp. nov.	–	–	–	–	–	–	×
Portunus obvallatus sp. nov.	–	–	–	×	×	–	–
Portunus woodwardi sp. nov.	–	–	–	×	×	–	–
Galene stipata sp. nov.	–	–	–	×	×	×	–
Parthenope (Rhinolambrus) sublitoralis sp. nov.	–	–	–	×	–	–	–
Ampliura simplex gen. et sp. nov.	–	–	–	–	–	×	–
Drachiella guinotae sp. nov.	–	–	–	×	–	–	–
Iphiculus granulatus sp. nov.	–	–	–	–	–	×	–
Iphiculus miriensis sp. nov.	–	–	–	×	×	–	–
Iphiculus sexspinosus sp. nov.	–	–	–	–	×	×	–
Leucosia longiangulata sp. nov.	–	–	–	×	–	–	–
Leucosia serenei sp. nov.	–	–	–	×	×	–	–
Leucosia tutongensis sp. nov.	–	–	–	×	×	–	–
Myra brevisulcata sp. nov.	–	–	–	×	–	–	–
Myra subcarinata sp. nov.	–	–	–	×	×	–	–
Myra trispinosa sp. nov.	–	–	–	×	×	–	–
Nucia borneoensis sp. nov.	–	–	–	×	–	–	–
Nucia calculoides sp. nov.	–	–	–	×	–	–	–
Nucia coxi sp. nov.	×	–	–	–	–	–	–
Nucilobus symmetricus gen. et sp. nov.	–	–	–	×	–	–	–
Pariphiculus gselli Beets *beetsi* subsp. nov.	–	–	–	×	–	–	–
Pariphiculus papillosus sp. nov.	–	–	–	×	–	–	–
Pariphiculus verrucosus sp. nov.	–	–	–	×	–	–	–
Philyra granulosa sp. nov.	–	–	–	×	×	–	–
Typilobus marginatus sp. nov.	–	×	–	–	–	–	–
Typilobus sp.	×	–	–	–	–	–	–
Palaeograpsus bittneri sp. nov.	–	–	–	–	×	–	–
Pinnixa aequipunctata sp. nov.	–	–	–	–	×	–	–
Pinnixa omega sp. nov.	–	–	–	–	×	–	–
Prepaeduma decapoda gen. et sp. nov.	–	–	–	×	×	–	–
Xenophthalmus subitus sp. nov.	–	–	–	–	×	–	–
Macrophthalmus (Mareotis) wilfordi sp. nov.	–	–	–	–	–	–	×

In 59005 (Fig. 70), In 59006–7, In 59008 (Fig. 71), In 59009–11 (about 50 specimens), In 61810 (Fig. 72), In 62120 (Figs 69a–c). All from locality S.4918, ? late Middle Pleistocene.

DESCRIPTION. The carapace is subquadrate in outline, widest between the tips of the outer orbital spines and with the regions distinctly defined by furrows. The moderately bilobed front is narrow, deflexed and constricted between the bases of the ocular peduncles; the front edge is smooth but the sides are finely granulate; there is a deep, narrow median furrow and a few surface granules. The upper orbital border is curved and slopes slightly backwards; the margin is lined with small, rounded granules; the lower orbital margin is studded along its entire length with longer, tubercular granules. The outer orbital spine is large, rectangular with granulated margins; it is directed outwards and slightly forwards and is separated from the 2nd lateral spine by a broad U-shaped notch. The 2nd lateral spine is somewhat weaker than the former and does not project beyond it, its anterior margin is convex and both that and the posterior one are weakly granulate; a shallow notch separates it from the small, triangular 3rd lateral spine.

With the exception of a narrow frontal strip and a few small central areas the dorsal surface is covered with coarse granules. On each metabranchial region two longitudinal rows of granules extend subparallel to the lateral margin; the inner row is 'broken' and becomes sigmoidal centrally. The lateral margins are subparallel and lined with granules.

The sides of the 4th and 5th somites of the male abdomen are nearly straight and parallel, while the sides of the 6th are slightly convex and taper a little distally.

The chelipeds associated with the carapaces are incomplete, and of those attributed to *M. wilfordi* none has a ridge on the outer margin of the palm, the upper part of the outer margin is granulate especially proximally, and the lower margin is smooth. The fixed finger is deflexed and one left-hand example has a large crenulated tooth proximally on the cutting edge, and the lower margin is granulate with the granules decreasing in size distally.

DISCUSSION. In having a narrow front with a well-developed ocular constriction, longitudinal granulate rows on the metabranchial lobes, a deflexed finger and, in the male, an abdomen with sides almost parallel, it would appear that the new species has already diverged considerably from the hypothetical ancestral form as envisaged by Barnes (1967: 250).

Macrophthalmus wilfordi has all the general characters essential to the subgenus *Mareotis* which one can reasonably expect to find preserved among fossil specimens. The occurrence of the greatest carapace width across the tips of the outer orbital spines instead of the 2nd anterolateral spines (common to the extant species of *Mareotis*) cannot be considered sufficient grounds to exclude it from that subgenus.

Of the nine Recent species placed by Barnes (1967, 1970) in *Mareotis*, *M. japonicus* de Haan compares favourably with and could well be descended from *M. wilfordi*. Apart from the position of the greatest carapace width, the former species may be distinguished by the upper and lower orbital margins being studded with similar-sized granules, by the absence of granules lining the frontal margin, and the presence of a transverse row of granules on the metabranchial lobes. According to Barnes (1967: 226) the inner longitudinal row of granules on the metabranchial lobes on Japanese forms of the Recent species is 'broken', as in *M. wilfordi*, while in Australian and North Chinese forms (Barnes 1970; 228) it is entire.

A superficial resemblance exists between *M. wilfordi* and *M. (Euplax) latreillei* (Desmarest, 1817), but the latter may be distinguished by the straight orbitofrontal margin, transverse rows of granules on the metabranchial lobes and a straight, undeflexed fixed finger.

Note. Barnes (1966), in revising the genus *Euplax* H. Milne Edwards, 1852 (type species *E. leptophthalmus* Milne Edwards, 1852, by subsequent designation of Barnes, 1966: 370) noted that *E. leptophthalmus* belonged to *Macrophthalmus* of the *M. latreillei* group. Later, Barnes (1967) erected a series of subgenera for the genus *Macrophthalmus*, including *M. (Venitus)* with type species by original designation *M. latreillei* (Desmarest, 1817). Thus it can be seen that *Euplax* is in fact a senior subjective synonym of *Venitus*. Further *Cyphoplax* Haime, 1855 (type species *Goneplax impressa* Desmarest, 1817 by monotypy, = *M. latreillei* (Barnes 1977: 280)) is also a senior subjective synonym of *Venitus*, but a junior subjective synonym of *Euplax*.

REFERENCES

Adams, A. & White, A. 1848. Crustacea. *In: The Zoology of the voyage of H.M.S. 'Samarang' 1843–1848*. 66 pp., 13 pls. London.

Alcock, A. W. 1896. Materials for a carcinological fauna of India, No. 2. The Brachyura Oxystomata. *Journal of the Asiatic Society of Bengal*, **65** (1): 134–296, pls 6–8.

—— 1900. The Brachyura Catometopa or Grapsoidea: Materials for a Carcinological Fauna of India. *Journal of the Asiatic Society of Bengal*, **69**: 279–456.

—— & Anderson, A. R. S. 1894. Natural History Notes from H.M. Indian Survey Steamer *Investigator*, Commander C. F. Oldham, R.N. Commanding. Series II, No. 14. An Account of Recent Collections of Deep Sea Crustacea from the Bay of Bengal and Laccadive Sea. *Journal of the Asiatic Society of Bengal*, **63** (3): 141–185, 9 pls.

Aubouin, J. 1965. *Geosynclines*. xv+335 pp. Amsterdam, &c. (Developments in Geotectonics **1**).

Audinet-Serville, J. G. 1839. *Histoire naturelle des Insectes. Orthoptères*. xviii + 776 pp., 14 pls. Paris.

Bachmayer, F. & Mohanti, M. 1973. Neue fossile Krebse aus dem Tertiär von Ost-Indien. *Annalen des Naturhistorischen Museums, Wien*, **77**: 63–67, 3 pls.

Barnes, R. S. K. 1966. The status of the crab genus *Euplax* H. Milne Edwards, 1852; and a new genus *Australoplax* of the subfamily Macrophthalminae Dana, 1851 (Brachyura: Ocypodidae). *Australian Zoologist*, Sydney, **13**: 370–376, pl. 24.

—— 1967. The Macrophthalminae of Australasia; with a review of the evolution and morphological diversity of the type genus *Macrophthalmus* (Crustacea: Brachyura). *Transactions of the Zoological Society of London*, **31**: 195–262, 4 pls.

—— 1970. The species of *Macrophthalmus* (Crustacea: Brachyura) in the Collection of the British Museum (Natural History). *Bulletin of the British Museum (Natural History)*, London, (Zoology) **20** (7): 205–251, 10 figs.

—— 1977. Concluding contribution towards a revision of, and a key to, the genus *Macrophthalmus* (Crustacea: Brachyura). *Journal of Zoology*, London, **182**: 267–280.

Beets, C. 1950. On fossil brachyuran crabs from the East Indies. *Verhandelingen van het Koninklijk Nederlandsch Geologisch-Mijnbouwkundig Genootschap*, (Geol. Ser.) **15**: 349–356, pl. 1.

Bell, T. 1855. A monograph of the Leucosiadae. *Transactions of the Linnean Society of London*, **21**: 277–314, pls 30–34.

—— 1859. Description of a New Genus of Crustacea, of the Family Pinnotheridae, in Which the Fifth Pair of Legs are Reduced to an Almost Imperceptible Rudiment. *Journal of the Proceeding of the Linnean Society*, London, (Zoology) **3**: 27–29.

Bemmelen, R. W. van 1970. *The Geology of Indonesia*, **1A**. *General Geology of Indonesia and Adjacent Archipelagoes*. 2nd edn. 732 pp. The Hague.

Bittner, A. 1875. Die Brachyuren des Vicentinischen Tertiärgebirges. *Denkschriften des Naturhistorischen Staatsmuseums, Wien*, **34**: 63–106, pls 1–5.

Böhm, J. 1922. *In*: Martin, K., Die Fossilien von Java. I Bd., 2. Abt. Arthropoda Crustacea. *Sammlung des Geologischen Reichsmuseums in Leiden*, (N.F.) **1**: 521–535, pl. 63.

Bol, A. J. & Hoorn, B. van 1980. Structural Styles in Western Sabah Offshore. *Bulletin Geological Society of Malaysia*, **12**: 1–16.

Cocco, A. 1832. Su di alcuni nuovi crustacei de'mari di Messina. Lettera del dott. Anastasio Cocco al celebre dott. William Elford Leach uno d'conservatori del Museo britannico in Londra. *Effemeridi scientifiche e letterarie per la Sicilia*, Palermo, **2**: 203–209, 1 pl.

Collenette, P. 1954. The Coal Deposits and a summary of the Geology of the Silimpopon Area, Tawau District, Colony of North Borneo. *Memoirs Geological Survey Department, British Territories in Borneo*, Kuching, **2**. 74 pp., 12 pls.

Dana, J. D. 1851. On the Classification of the Cancroidea. *American Journal of Science and Arts*, New Haven, (2) **12**: 121–131.

—— 1852. Crustacea, Part I. *In*: *United States Exploring Expedition during the Years 1838, 1839, 1840, 1841, 1842 under the Command of Captain Charles Wilkes, U.S.N.*, Philadelphia, **13**. 685 pp.

Desmarest, A. G. 1817. *Crustacés Fossiles. In*: Biot, J. B. *et al., Nouveau Dictionnaire d'Histoire Naturelle . . .* (Nouvelle edition), **8**: 495–519. Paris.

Edmondson, C. H. 1954. Hawaiian Portunidae. *Occasional Papers of the Bernice Pauahi Bishop Museum*, Honolulu, **21** (12): 217–274, 44 figs.

Fabricius, J. C. 1775. *Systema entomologicae, sistens insectorum classes, ordines, genera, species, adiectis synonymis, locis descriptionibus, observationibus*. 832 pp. Flensburg & Lipsiae.

—— 1793. *J. C. Fabricii . . . Entomologia systematica emendata et aucta . . . adjectis synonymis, locis, observationibus descriptionibus*. **2**. viii + 519 pp. Hafniae.

—— 1798. *Supplementum Entomologiae systematicae*. 572 pp. Hafniae.

Fitch, F. H. 1958. The geology and mineral resources of the Sandakan area and parts of the Kinabalangan and Labuk Valley, North Borneo. *Memoirs Geological Survey Department, British Territories in Borneo*, Kuching, **9**. 202 pp.

Glaessner, M. F. 1929. *In*: Pompeckj, F. J. (ed.), *Fossilium Catalogus* I: Animalia, **41** (Crustacea, Decapoda). 464 pp. Berlin.

Gómez-Alba, J. A. S. 1988. *Guía de Campo de los Fósiles de España y de Europa*. xliv + 925 pp., 388 pls. Barcelona (Ediciones Omega).

Guinot, D. 1977. Propositions pour une nouvelle classification des Crustacés Décapodes Brachyoures. *Compte Rendu Hebdomaire des Séances de l'Académie des Sciences Paris*, (D) **285**: 1049–1052.

—— 1979. Données nouvelles sur la morphologie, la phylogenèse et la taxonomie des Crustacés Décapodes Brachyoures. *Mémoires du Muséum National d'Histoire Naturelle*, Paris, (A) **112**. 354 pp., 27 pls.

Haan, W. de 1833–50. Crustacea. *In*: Siebold, P. F. von, *Fauna Japonica, sive descriptio animalium, quae in itinere per Japoniam, jussu et auspiciis superiorum, qui summum in India, Batava Imperium tenent, suscepto, annis 1823–1830 collegit, notis observationibus et adumbrationibus illustravit*. ix–xvi, 1–xxxi, vii–xvii + 243 pp., pls A–J, L–Q, 1–55, circ. 2. Lugduni Batavorum [Leiden].

Haile, N. S. 1969. Geosynclinal theory and the organizational pattern of the North-west Borneo Geosyncline. *Quarterly Journal of the Geological Society of London*, **124**: 171–195.

—— & Wong, N. P. Y. 1965. The geology and mineral resources of Dent Peninsula, Sabah. *Memoir Geological Survey Borneo Region, Malaysia*, Kuching, **16**. 199 pp.

Haime, J. 1855. Notice sur la géologie de l'île Majorque. *Bulletin de la Société Geologique de France*, Paris, (2) **12**: 734–752, pl. 15.

Haswell, W. A. 1880. Contributions to a Monograph of Australian Leucosiidae. *Proceedings of the Linnaean Society of New South Wales*, Sydney, **4**: 44–60, pls 5, 6.

Henderson, T. R. 1893. A contribution to Indian carcinology. *Transactions of the Linnean Society of London*, (Zoology) (2) **5**: 325–458, pls 36–40.

Herbst, J. F. W. 1782–1804. *Versuch einer Naturgeschichte der Krabben und Krebse*. 3 vols. 274 + 226 (216) pp., 72 pls. Berlin & Stralsund.

Holthuis, L. B. 1959. Notes on pre-Linnean Carcinology (Including the study of Xiphosura). *In*: Wit, H. C. D. de (ed.), *Rumphius Memorial Volume*: 63–125, photos 7–11. Baarn.

Ihle, J. E. W. 1918. Die Decapoda Brachyura der Siboga-Expedition. III. Oxystomata, Calappidae, Leucosiidae, Raninidae. *Siboga-Expeditie*, **39** (62): 159–322, figs 78–148.

Jacquinot, H. & Lucas, H. 1853. *In*: Hombron, J. B. & Jacquinot, H. (eds), *Voyage au Pôle Sud et dans l'Océanie . . . exécuté . . . pendant 1837–40*. **3**, Crustacés. 103 pp., 9 pls. Paris.

Lamarck, J. B. P. A. de 1801. *Systême des Animaux sans vertèbres . . . précédé du discours d'ouverture du Cours de Zoologie donné dans le Muséum National d'Histoire Naturelle, l'an 8*. viii + 432 pp. Paris.

Lankester, E. R. 1904. The Structure and Classification of the Arthropoda. *Quarterly Journal of Microscopical Science*, London, **47**: 523–576, pl. 42.

Latreille, P. A. 1810. *Considérations générales sur l'ordre naturel des animaux composant les classes des Crustacés, des Arachnides, et des Insectes, avec un tableau méthodique de leurs genres, disposés en familles*. 444 pp. Paris.

—— 1817. Les crustacés, les arachnides et les insectes. *In*: Cuvier, G., *Le règne animal distribué d'après son organization, pour servir de base à l'histoire naturelle des animaux et d'introduction à l'anatomie comparée*. **3**. xxix + 653 pp. Paris.

—— 1823. *In*: Desmarest, A.-G., Malacostracés, Malacostraca. *Dictionnaire des Sciences Naturelles* **28**: 138–425. Strasburg & Paris.

—— 1825. *Familles naturelles du règne animal, exposées succinctement et dans un ordre analytique, avec l'indication de leurs genres*. 570 pp. Paris.

Laurie, R. D. 1906. Report on the Brachyura, collected by Prof. Herdman at Ceylon, 1902. *Report Ceylon Pearl Oyster Fisheries and Marine Biology*, Part V, **5**: 349–432, pls 1, 2.

Leach, W. E. 1817. Monograph on the genera and species of the Malacostracous fam. Leucosidae. *In*: *The Zoological miscellany being description of new and interesting animals*, **3**: 17–26. London.

Liechti, P. 1960. The geology of Sarawak, Brunei and the Western part of North Borneo. *Bulletin Geological Survey Department, British Territories in Borneo*, **3**. 360 pp.

Linnaeus, C. 1758. *Systema Naturae*. 10th edn, **1**. 824 pp. Holmiae.

—— 1767. *Systema Naturae . . . 12th edn, 1*, Regnum Animale (2): 533–1327. Holmiae.

Macleay, W. S. 1838. *Illustrations of the Zoology of South Africa; Annulosa*. 75 pp., 4 pls. London.

Manning, R. B. & Holthuis, L. B. 1981. West African Brachyuran Crabs (Crustacea: Decapoda). *Smithsonian Contributions to Zoology*, Washington, **306**. 379 pp.

Miers, E. J. 1886. Report on the Brachyura Collected by H.M.S. *Challenger* during the Years 1873–1876. *Report on the Scientific Results of the Voyage of H.M.S. Challenger during the Years 1873–76, Zoology*, **17**. xli + 363 pp., 29 pls.

Milne Edwards, A. 1861. Études zoologiques sur les Crustacés Récents de la Famille des Portuniens. *Archives du Muséum d'Histoire Naturelle*, Paris, **10**: 319–430, pls 28–38.

—— 1865. Monographie des Crustacés fossiles de la famille des Cancériens. *Annales des Sciences Naturelles, Zoology*, Paris, (5) **3**: 297–351, pls 5–13.

—— 1874. Recherches sur la faune carcinologique de la Nouvelle-Calédonie. *Nouvelles Archives du Muséum d'Histoire Naturelle, Paris*, **10**: 39–58, pls 2, 3.

—— 1878. *Études sur les Xiphosures et sur les Podophthalmiens. Mission scientifique au Méxique*, **5** (8): 121–184, pls 21–27, 29, 30. Paris.

Milne Edwards, H. 1834–40. *Histoire naturelle des Crustacés*. I: xxxv + 468 pp. II: 532 pp. III: 638 pp. Atlas. Paris.

—— 1840–41. *In*: *Le Règne Animal distribué d'après son organisation, pour servir de base à l'histoire naturelle des animaux et d'introduction à l'anatomie comparée. Edition accompagnée de planches gravées, représentant les types de tous les genres, les caractères distinctifs des divers groupes et les modifications de structure sur lesquelles repose cette classification; par Une Réunion de Disciples de Cuvier*, Crustacés Livr. 99, pl. 10 (July 1840); *loc. cit.* Livr. 120, pl. 16 (May 1841).

—— 1852. Observations sur les affinités zoologiques et la classification naturelle des crustacés. *Annales des Sciences Naturelles, Zoology*, Paris, (3) **18**: 109–166, pls 3, 4.

Nuttall, C. P. 1961. Gastropoda from the Miri and Seria Formations, Tutong Road, Brunei. *In*: Wilford, G. E., The Geology and Mineral Resources of

Brunei and adjacent parts of Sarawak with descriptions of Seria and Miri Oilfields. *Memoir Geological Survey Department, British Territories in Borneo*, Brunei, **10**. 319 pp., 77 pls, 3 maps.

Quayle, J. & Collins, J. S. H. 1981. New Eocene crabs from the Hampshire Basin. *Palaeontology*, London, **24**: 733–758, pls 104, 105.

Rafinesque-Schmaltz, C. S. 1815. *Analyse de la nature, ou tableau de l'univers et des corps organisés*. 224 pp. Palermo.

Rathbun, M. J. 1897. A Revision of the Nomenclature of Brachyura. *Proceedings of the Biological Society of Washington*, **11**: 149–151.

—— 1904. Some changes in crustacean nomenclature. *Proceedings of the Biological Society of Washington*, **17**: 169–172.

—— 1909. New Crabs from the Gulf of Siam. *Proceedings of the Biological Society of Washington*, **22**: 107–114.

—— 1910. The Danish Expedition to Siam, 1899–1900. V. Brachyura. *Kongelige Danske Videnskabernes Selskabs Skrifter*, Copenhagen, (7) **5** (4): 301–368, 2 pls.

—— 1925. The Spider Crabs of America. *Bulletin of the United States National Museum*, Washington, D.C., **129**. 613 pp., 283 pls.

—— 1926. Fossil stalk-eyed crustacea of the Pacific slope of North America. *Bulletin of the United States National Museum*, Washington, **138**. 155 pp., 39 pls.

—— 1932. Fossil Pinnotherids from the California Miocene. *Journal of the Washington Academy of Sciences*, **22** (14): 411–413.

Rüppell, W. P. E. S. 1830. *Beschreibung und Abbildung von 24 Arten Kurzschwanigen Krabben als Beiträge zur Naturgeschichte des Rothen Meeres*. 28 pp., 6 pls. Frankfurt.

Sakai, T. 1934. Brachyura from the coast of Kyusyu, Japan. *Science Reports of the Tokyo Bunrika Daigaku*, (B) **1**: 281–330, 2 pls.

—— 1965. *The Crabs of Sagami Bay*. xvi + 206 pp., 100 pls. Biological Laboratories of the Imperial Household, Tokyo.

Samouelle, G. 1819. *The Entomologist's Useful Compendium, or an Introduction to the Knowledge of British Insects*. 496 pp. London.

Say, T. 1818. Appendix to the Account of the Crustacea of the United States. *Journal of the Academy of Natural Sciences of Philadelphia*, **1** (16): 445–458.

Schuppli, H. M. 1946. Geology of oil basins of East Indian Archipelago. *Bulletin of the American Association of Petroleum Geologists*, Chicago, **30**: 1–22.

Serène, R. 1968. The Brachyura of the Indo-West Pacific region. *In*: *Prodromus for a Check List of the (non-planctonic) Marine Fauna of Southeast Asia*. UNESCO Singapore, Spec. Publ. 1 (Fauna III C$_{c3}$): 33–112 (Roneotyped).

—— & Soh, C. L. 1976. Brachyura collected during the Thai-Danish Expedition (1966). *Research Bulletin Phuket Marine Biological Center*, **12**. 52 pp., 8 pls.

Stimpson, W. 1858. Prodromus descriptionis animalium evertebratorum, quae in Expeditione ad Oceanum Pacificum Septentrionalem, a Republica Federata missa, Cadwaladaro Ringgold et Johanne Rodgers Ducibus, observavit et descripsit. *Proceedings of the Academy of Natural Sciences of Philadelphia*, **1857**: 159–163.

Stoliczka, F. 1871. Observations on fossil crabs from Tertiary deposits in Sind and Kutch. *Memoirs of the Geological Survey of India. Palaeontologia Indica*, Calcutta, (7) **14** [vol. 1, pt 1]. 16 pp., 5 pls.

Straelen, V. van 1923. Note sur la position systematique de quelques Crustacés décapodes de l'époque crétacée. *Bulletins de la Classe des Sciences, Académie Royale de Belgique*, Bruxelles, (5) **9**: 116–125.

Umbgrove, J. H. F. 1933. Verschillende typen van tertaire geosynclinalen in den Indischen Archipel. *Leidsche Geologische Mededeelingen*, **6**: 33–43.

Via, L. 1959. Decápodos fósiles del Eoceno español. *Boletin del Instituto Geológico y Minero de España*, Madrid, **70**: 331–402, 7 pls.

Via Boada, L. 1969. Crustáceos decápodos del Eoceno español. *Pirineos*, Jaca, **91–94**. 479 pp., 39 pls.

Weber, F. 1795. *Nomenclator Entomologicus secundum Entomologiam systematicam ill. Fabricii, adjectis speciebus recens detectis et varietatibus*. viii + 171 pp. Chilonii [Kiel] & Hamburgi.

Wenk, E. 1938. *Report on a rapid geological reconnaissance of southeastern British North Borneo and on resulting stratigraphical correlations*. Confidential report of Sarawak Oilfields Ltd.

White, A. W. 1846. Notes on four new Genera of Crustacea. *Annals and Magazine of Natural History*, London, **18**: 176–178, pl. 2, figs 1–3.

Wilford, G. E. 1961. The Geology and Mineral Resources of Brunei and adjacent parts of Sarawak with descriptions of Seria and Miri Oilfields. *Memoir Geological Survey Department, British Territories in Borneo*, Brunei, **10**. 319 pp., 77 pls, 3 maps.

Woodward, H. 1905. Note on a Fossil Crab and a Group of Balani discovered in Concretions on the Beach at Ormara Headland, Mekran Coast. *Geological Magazine*, London, (5) **2**: 305–310, figs 1, 2.

Bull. Br. Mus. nat. Hist. (Geol.) **47** (1): 35–50
Issued 29 August 1991

Two new pseudosciurids (Rodentia, Mammalia) from the English Late Eocene, and their implications for phylogeny and speciation

J. J. HOOKER

Department of Palaeontology, Natural History Museum, Cromwell Road, London SW7 5BD

SYNOPSIS. Assemblages formerly referred to *Treposciurus intermedius* and *Suevosciurus palustris*, from the Solent Group (Late Eocene) of Hordle and of localities in the Isle of Wight (Hampshire Basin), are shown to differ markedly from the type specimens of these species. They are here described as two new species belonging to the original genera. The differences between these two superficially similar species are clarified and evidence for phylogeny and speciation events within the genera is discussed.

INTRODUCTION

Bosma (1974), when describing the rodent faunas of the Isle of Wight Late Eocene and Early Oligocene, attributed two small pseudosciurid species to *Treposciurus intermedius* (Schlosser 1884) and *Suevosciurus palustris* (Misonne 1957). The lectotype of the former is a dentary from the Phosphorites du Quercy of Escamps (old locality), Tarn, southern France, of imprecise but probable Late Eocene age. The holotype of the latter is an upper M^1 or M^2 from the Sables de Boutersem (Early Oligocene – i.e. immediately post-Grande Coupure) of Hoogbutsel, Belgium. Both type specimens are thus geographically, and in at least one case also stratigraphically, distant from the southern English referred material. They also differ from them in both size and morphology. The *Treposciurus* is rare, but new material from the Hordle Mammal Bed makes it better known.

Abbreviations

The following relate to institutes and/or their specimen numbers. BSPG = Bayerische Staatssammlung für Paläontologie und historische Geologie, München; GIU = Instituut voor Aardwetenschappen, Rijksuniversitet Utrecht; IRSNB = Institut Royal des Sciences Naturelles de Belgique, Bruxelles; M = register numbers of the Mammal Section, Department of Palaeontology, Natural History Museum, London.

Synonymies

Procedure and terminology follow Matthews (1973).

SYSTEMATIC DESCRIPTIONS

Order **RODENTIA**
Superfamily **THERIDOMYOIDEA**
Family **PSEUDOSCIURIDAE**

Genus *TREPOSCIURUS* Schmidt-Kittler 1970

TYPE SPECIES. *Treposciurus mutabilis* Schmidt-Kittler 1970; Late Eocene, Bavaria, southern Germany.

Treposciurus gardneri sp. nov.   Figs 1–14

?vp	1973	*Suevosciurus* (*Microsuevosciurus*) aff. *minimus* (Major 1873); Hartenberger: 16; pl. 1, figs 7–9, 11–13.
vp.	1974	*Treposciurus intermedius* (Schlosser 1884); Bosma: 49–52; pl. 6, figs 3–10.
vp.	1974	*Suevosciurus palustris* (Misonne 1957); Bosma: pl. 5, fig. 7.
v.	*1980*	*Treposciurus intermedius* (Schlosser 1884); Hooker & Insole: 39.
v.	*1982*	*Treposciurus intermedius* (Schlosser 1884); Russell *et al.*: 57.
v.	1986	*Treposciurus intermedius* (Schlosser 1884); Hooker: 308–311.
v.	*1987*	*Treposciurus intermedius* (Schlosser 1884); Collinson & Hooker: 292.
v.	1987	*Treposciurus* sp. nov.; Hooker: 112.
v.	1989	*Treposciurus* sp. nov.; Hooker: fig. 2.

HOLOTYPE. Right maxilla with DP^4, M^{1-3} (M44472).

PARATYPES. DP^4 (M44473), 7 $M^{1/2}$ (M44474–80), 3 DP_4 (M44481–3), 2 P_4 (M44484–5), right dentary fragment with worn M_{1-2} (M44486), 9 $M_{1/2}$ (M44487–95), 4 M_3 (M44496–9).

NAME. After Mr R.G. Gardner who collected the type series.

TYPE HORIZON AND LOCALITY. Mammal Bed (see Cray 1973), Totland Bay Member, Headon Hill Formation (see Insole & Daley 1985; previously informally referred to as Lower Headon Beds), Hordle, Hampshire.

REFERRED MATERIAL. Isolated teeth described and figured by Bosma (1974) from sample localities HH1 and HH2, Totland Bay Member; HH3 and other sample localities from 'below the main lignite band', Hatherwood Limestone Member – two M^3s and an $M_{1/2}$ from HH3 (GIU 492, 412 and 499) and an $M^{1/2}$ and an $M_{1/2}$ from HH4 (GIU 480 and 426), referred by Bosma to *Suevosciurus palustris* also belong here; all Headon Hill Formation, Headon Hill, Isle of Wight. Also from sample locality WB2A, Bembridge Marls Member, Bouldnor Formation, Whitecliff Bay, Isle of Wight. An $M^{1/2}$ (M51083) from a dark clay at top of limestone overlying *Cyrena cycladiformis* bed (Bristow *et al.* 1889), Totland Bay Member; a DP^4 (M51084) from shelly lenses at base of lignite bed, Hatherwood Limestone Member (including

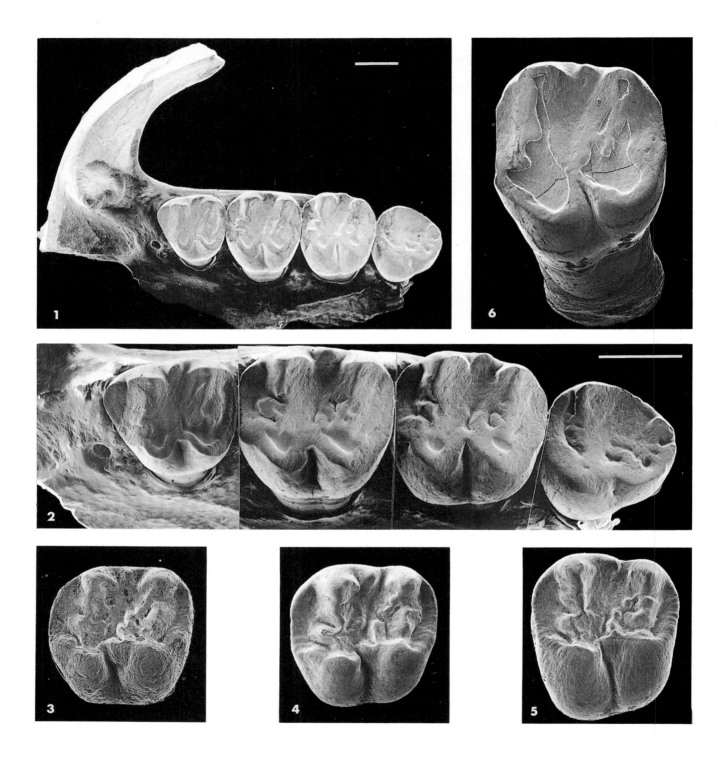

Figs 1–6 Scanning electron micrographs (SEMs) of maxilla and upper cheek teeth in occlusal view of *Treposciurus gardneri* sp. nov., Solent Group, late Eocene. Figs 1–5, Mammal Bed, Totland Bay Member, Headon Hill Formation, Hordle, Hants. Figs 1–2, **holotype**, right maxilla, with DP^4, M^{1-3} and part of zygomatic arch (reversed) (M44472); 2 shows details of the teeth. Figs 3–5, paratype right $M^{1/2}$s (reversed) (M44474–6). Fig. 6, Bembridge Limestone Formation, Headon Hill, Isle of Wight. Referred right $M^{1/2}$ (reversed) (M51085). Scale bars = 1 mm; specimens uncoated.

sample locality of HH3); isolated upper and lower M1/2s (M44500, M51085–7, M51104) from argillaceous beds within the Bembridge Limestone Formation (including HH6–7 of Bosma, 1974); all Headon Hill; and an M_3 from above a calcareous sandstone, Bembridge Marls Member (level of WB2 of Bosma, 1974), Whitecliff Bay. Probably also the isolated cheek teeth described and figured by Hartenberger (1973) from the upper Calcaire de Fons, Fons 4, Gard, France, as *Suevosciurus* (*Microsuevosciurus*) aff. *minimus*.

DIAGNOSIS. Small species of *Treposciurus* (mean length of $M^{1/2}$ 1·77 mm); cheek teeth with relatively shallow basins, without dense enamel wrinkling or reticulation; upper pre-ultimate molars and DP^4 with dentine cored (visible only after heavy wear – Fig. 6), interrupted metalophule I, usually extending buccally to the metacone and joining the endoloph between the hypocone and sinus; lower cheek teeth with weak mesoconid protruding only a short distance into sinusid; lower molars with medium-sized anteroconid, joined to interrupted metalophulid by anterolophulid; M_{1-2} mesial hypoconid wall more or less vertical, encroaching little on sinusid, which consequently has open appearance; buccal corners of upper molars often noticeably rounded, reflecting lingual retreat of anteroloph and posteroloph.

DIFFERENTIAL DIAGNOSIS. Other *Treposciurus* species are larger, tend to have densely wrinkled enamel and/or a lesser development of M^{1-2} metalophule I and have lower molars with stronger mesoconids and weaker anteroconids (see Schmidt-Kittler 1971; Bosma 1974; Hooker 1986).

DESCRIPTION. P_4, DP_4 and $M_{1/2}$ tooth types are made known for the first time and contribute significantly to an understanding of the species and its distinction from *T. intermedius* (see p. 39). Variation in the separation of the protoconid and metaconid on the two P_4s can be seen in Figs 11–12. Complex variation in development of M^{1-2} metalophule I and of the coarse enamel folds in upper and lower molars can be seen in Figs 1–14 and Bosma (1974: pl. 6, figs 3–10). Other features such as strength of the lower cheek tooth mesoconid and anteroconid are remarkably constant, within the relatively small sample. The rounding of the buccal corners of upper molars is striking when present, but by no means constant. It is most marked in the specimens from WB2A (Bosma 1974: pl. 6, figs 6, 8–10). The outline shape of M_3 varies much (Figs 10, 14) in a similar way to that documented for *Treposciurus helveticus preecei* from the Bartonian of Creechbarrow (Hooker 1986: pl. 17, figs 4, 6), and reflects differing development of the entoconid and transverseness versus obliquity of the mesial margin. The length measurements of upper and lower preultimate molars have a low coefficient of variation (see Table 1). The holotype maxilla is broken a short distance anteriorly and medially of DP^4 so the extent of the incisive foramen is unknown.

Other assemblages differ from the type assemblage slightly in size, but not in morphology (Table 1). The teeth from Fons 4 have mean lengths of $M^{1/2}$ (1·62 mm) and $M_{1/2}$ (1·65 mm) (Hartenberger 1973:16) slightly less than has the type assemblage of *T. gardneri*, with little overlap of measurements. Morphology clearly shows the Fons 4 assemblage is very

Table 1 Statistics of length and width measurements of cheek teeth of *Treposciurus*. (N = number of specimens; OR = observed range; x̄ = mean; s = standard deviation; v = coefficient of variation. Measurement in brackets is estimate). * = measurements taken from epoxy casts.

Sp./loc.	Tooth	N	OR	x̄	s	v	N	OR	x̄	s	v
				Length					Width		
T. gardneri	DP^4	2	1·68–1·88	1·78			2	1·52–1·68	1·60		
Hordle	$M^{1/2}$	8	1·66–1·80	1·77	0·048	2·70	8	1·61–2·04	1·82	0·143	7·85
Mammal Bed	M^3	1		1·60			1		1·76		
	DP_4	3	1·72–1·98	1·82			3	1·24–1·38	1·29		
	P_4	2	1·76–1·80	1·78			2	1·44–1·54	1·49		
	$M_{1/2}$	10	1·76–1·96	1·88	0·065	3·42	10	1·50–1·72	1·63	0·066	4·09
	M_3	4	1·86–2·02	1·94			3	1·52–1·66	1·58		
T. gardneri above Cyrena cycladiformis Bed	$M^{1/2}$	1		1·90			1		(1·96)		
T. gardneri HH3–4	DP^4	1		1·92			1		1·66		
	*$M^{1/2}$	1		1·65			1		1·64		
	*M^3	2	1·56–1·67	1·62			2	1·53–1·64	1·59		
	*$M_{1/2}$	2	1·74–1·86	1·80			2	1·45–1·56	1·51		
T. gardneri Bembridge Limestone	$M^{1/2}$	1		2·10			1		2·38		
	$M_{1/2}$	1		2·14			1		1·84		
*Lectotype *T. intermedius* Quercy	DP_4	1		2·48			1		1·80		
	M_1	1		2·56			1		2·24		
	M_2	1		2·72			1		2·26		
	M_3	1		2·60			1		1·92		

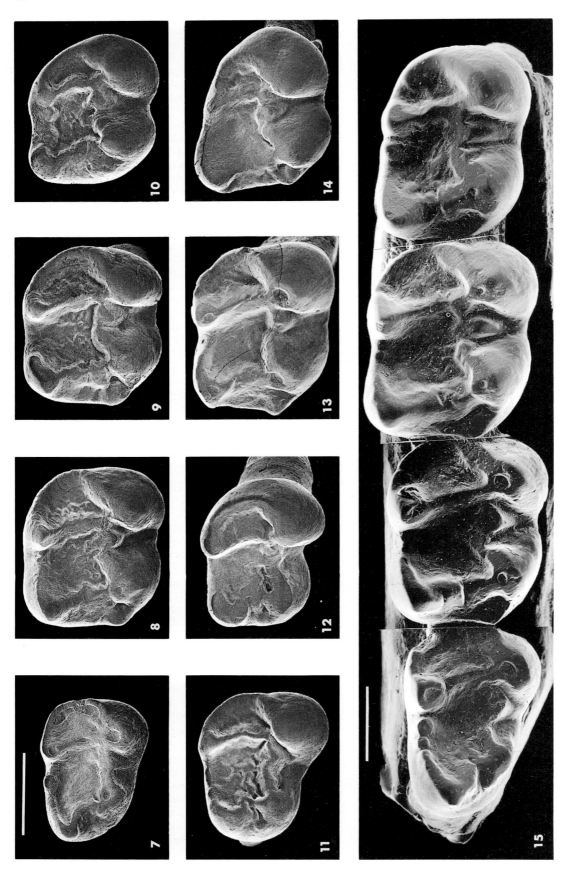

Figs 7–15 SEMs of lower cheek teeth in occlusal view of *Treposciurus*, late Eocene. Figs 7–12, 14, paratypes of *Treposciurus gardneri* sp. nov., from the Mammal Bed, Hordle, Hants. Fig. 7, right DP$_4$ (reversed) (M44481); Fig. 8, right M$_{1/2}$ (reversed) (M44487); Fig. 9, left M$_{1/2}$ (M44488); Fig. 10, left M$_3$ (M44496); Figs 11–12, left P$_4$s (M44484–5); Fig. 14, right M$_3$ (reversed) (M44497). Fig. 13, left M$_{1/2}$ of *T. gardneri* sp. nov. from the Bembridge Limestone Formation, Headon Hill, Isle of Wight (M51104). Fig. 15, epoxy cast (coated with gold-palladium) of right DP$_4$, M$_{1-3}$ of lectotype of *T. intermedius*, from the Phosphorites du Quercy, France (original BSPG.1879XV-192). Scale bars = 1 mm; original specimens uncoated.

closely related and, as Bosma (1974:53) noted, not a *Suevosciurus*, but the distinct plots of measurements leave doubt over precise conspecificity. The $M^{1/2}$s from WB2A described by Bosma also lie outside the measurements of the type assemblage, being slightly larger. The small number of specimens (3) suggests that there would be overlap if more were known, but the single $M^{1/2}$ from the Bembridge Limestone is also slightly larger than any from the type assemblage. The Bembridge Limestone Formation and the Bembridge Marls Member of the Bouldnor Formation are demonstrably younger than the Mammal Bed at Hordle, by superposition, and Fons 4 is considered older, e.g. according to the evolutionary grade of its *Choeropotamus* (see Sudre 1978). It would therefore seem that there is a trend towards size increase, which could indicate that the Fons 4, Hordle and WB2 assemblages are segments of a single lineage. Curiously, however, the two upper molars from HH1 recorded by Bosma (1974) are also larger than any in the type assemblage. As the Mammal Bed and HH1 (only 8·5 km apart) both contain *Thalerimys headonensis*, they are penecontemporaneous (belonging to Bosma's '*Isoptychus*' *headonensis* Zone). Either none of the assemblages has yet been adequately sampled to show the size range, or a more complex pattern of size than a simple increase through time may pertain.

Discussion

Explanation of reidentification.

The type series of '*Sciuroides*' *intermedius* Schlosser, 1884 consisted of two dentaries and a maxilla from the old Quercy Phosphorites locality of Escamps, southern France. Dehm (1937) selected one of the dentaries as lectotype. Schmidt-Kittler (1970) erected the genus *Treposciurus* for *T. mutabilis* Schmidt-Kittler 1970 (type species) and for '*Sciuroides*' *intermedius* Schlosser 1884. Schmidt-Kittler (1971) removed the maxilla of the type series of *T. intermedius* from that species, because the posterior border of its incisive foramen reached back to P^4 in Schlosser's figure, thus contrasting with another Quercy maxilla figured by Thaler (1966), the dental match of which was better. Bosma (1974) claimed that the posterior border of the incisive foramen was damaged in the paralectotype maxilla and that the specimen could still belong in *T. intermedius*. Bosma was relying on referred maxillary specimens from Quercy for comparison with her Isle of Wight material, as her knowledge of the lower dentition was restricted to a single M_3. She noted that these upper molars and the rather undiagnostic M_3 were morphologically similar to those of *T. intermedius*, but that they were slightly smaller. Whatever the true identity of the paralectotype maxilla, the lectotype dentary is all that one can rely upon for potential identification of other *T. intermedius* specimens. Inaccuracy of old Quercy locality names makes it impossible to recognize any unequivocal topotypes. Nevertheless, several upper dentitions from old Quercy collections appear to match adequately the lowers of the lectotype and other specimens; their distinguishing features have been tabulated by Schmidt-Kittler (1971: tab. 4). They include upper molars which differ strikingly from *T. gardneri* in having a symmetrical endoloph, coarsely wrinkled enamel and no metalophule I.

It can be seen that the teeth of the lectotype of *T.*

intermedius are significantly larger than the equivalents in *T. gardneri* (Figs 7–15). They have a stronger mesoconid and a higher length/width ratio. The M_{1-2} hypoconid encroaches further mesially on the sinusid and DP_4 and M_3 each have a prominent mesostylid. The last feature is variable in *T. gardneri*, occurring in one out of three DP_4s and incipiently in one out of ten $M_{1/2}$s; the others are constant for the available specimens.

What makes *T. gardneri* a *Treposciurus* and not a *Suevosciurus*? The recognition that the tooth types previously regarded as P^4_4 are in reality DP^4_4 (Hooker 1986) removes one distinction previously maintained. The most recently emended diagnoses of *Treposciurus* and *Suevosciurus* (Hooker 1986: 308, 315) do not polarize the characters, but a restriction to those in the advanced state is shown in Fig. 16. In *Suevosciurus*, upper and lower cheek teeth have basins that are deeper and more concave, the antero- and posterolophs and lophids tend to be more prominent and the transverse lophs are more distinct; the upper molars tend to have an uninterrupted, more symmetrical endoloph, with any expression of metaconule 2 or metalophule I (which lack a dentine core) restricted to a lingual section; metalophule II joins the hypocone; M^3 shows more distal reduction with, in the contemporaneous species from Hordle, total absence of metalophule II, a much shallower sinus and more mesially positioned protocone (compare the present figures with Bosma 1974: pl. 6, fig. 10). The lower molars usually lack an anterolophulid and have a stronger mesoconid which, in the contemporaneous species from Hordle (amongst others), is usually linked by a crest to the mesiobuccal corner of the hypoconid (Figs 22–23); and DP^4 usually has a concave mesiolingual margin (Figs 17–18).

Evidence for a late Eocene speciation event.

Hooker (1986) formulated a model of cladogenetic speciation in the genus *Treposciurus*. He envisaged a morphologically very variable *T. helveticus* of the Bartonian giving rise around the Bartonian–Ludian boundary to *T. intermedius* and *T. mutabilis* by respective selection of two morphs present together in the ancestral species and by size differentiation. He divided *T. helveticus* (raised to species level from *T. mutabilis helveticus* Schmidt-Kittler 1971) into two subspecies: a nominate one from Eclépens B, Switzerland, and *T. h. preecei* from Creechbarrow. Current detailed study of the Eclépens B material (Hooker & Weidmann, in prep.) shows that *T. h. helveticus* involves greater complexity. The Eclépens B *Treposciurus* is therefore simply referred to as *Treposciurus helveticus* Schmidt-Kittler 1971, and the subspecies *T. helveticus preecei* Hooker 1986 is hereby raised to species level as *Treposciurus preecei* Hooker 1986 new rank.

These nomenclatural changes do not affect the evidence for the speciation event, but they do mean that the ancestral species is now *T. preecei*, and one daughter species is *T. gardneri*, whilst the other daughter branch is at present an unresolved complex comprising *T. mutabilis*, *T. helveticus* and *T. intermedius*. The evidence based on character analysis is presented in Fig. 16. Here a cladogram shows the splitting of *Treposciurus gardneri* from the rest and also the characters linking the genus *Treposciurus* to its nearest sister taxon *Suevosciurus*. Character polarity was obtained by outgroup comparison with the rest of the Pseudosciuridae.

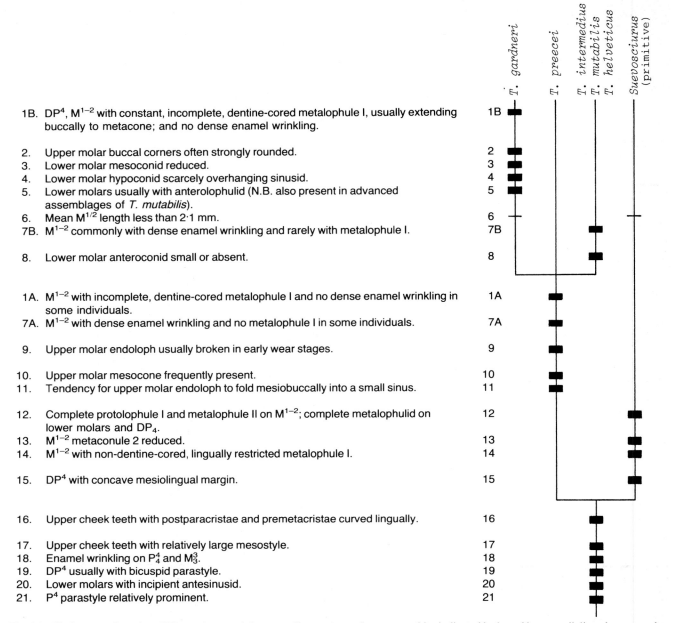

1B. DP⁴, M¹⁻² with constant, incomplete, dentine-cored metalophule I, usually extending buccally to metacone; and no dense enamel wrinkling.

2. Upper molar buccal corners often strongly rounded.
3. Lower molar mesoconid reduced.
4. Lower molar hypoconid scarcely overhanging sinusid.
5. Lower molars usually with anterolophulid (N.B. also present in advanced assemblages of *T. mutabilis*).
6. Mean M¹/² length less than 2·1 mm.
7B. M¹⁻² commonly with dense enamel wrinkling and rarely with metalophule I.

8. Lower molar anteroconid small or absent.

1A. M¹⁻² with incomplete, dentine-cored metalophule I and no dense enamel wrinkling in some individuals.
7A. M¹⁻² with dense enamel wrinkling and no metalophule I in some individuals.

9. Upper molar endoloph usually broken in early wear stages.

10. Upper molar mesocone frequently present.
11. Tendency for upper molar endoloph to fold mesiobuccally into a small sinus.

12. Complete protolophule I and metalophule II on M¹⁻²; complete metalophulid on lower molars and DP₄.
13. M¹⁻² metaconule 2 reduced.
14. M¹⁻² with non-dentine-cored, lingually restricted metalophule I.

15. DP⁴ with concave mesiolingual margin.

16. Upper cheek teeth with postparacristae and premetacristae curved lingually.

17. Upper cheek teeth with relatively large mesostyle.
18. Enamel wrinkling on P₄⁴ and M₃³.
19. DP⁴ usually with bicuspid parastyle.
20. Lower molars with incipient antesinusid.
21. P⁴ parastyle relatively prominent.

Fig. 16 Cladogram of species of *Treposciurus* and the genus *Suevosciurus*. Synapomorphies indicated by broad bar, parallelisms by narrow bar. Characters 1 and 7 are multistate.

Genus **SUEVOSCIURUS** Dehm 1937

Type species. *Sciuroides fraasi* Major 1873.

Suevosciurus bosmae sp. nov.   Figs 17–23, 30E, F

vp.	1974	*Suevosciurus palustris* (Misonne 1957); Bosma: 34–44; pl. 5, figs 7–9.
v.	1980	*Suevosciurus palustris* (Misonne 1957); Hooker & Insole: 39.
vp.	1982	*Suevosciurus palustris* (Misonne 1957); Russell *et al*.: 57.
v.	1986	*Suevosciurus*; Hooker: 322–327.

v.	1987	*Suevosciurus* sp. 1; Collinson & Hooker: 292.
v.	1987	*Suevosciurus* sp. nov.; Hooker: 112.
v.	1989	*Suevosciurus* sp. nov.; Hooker: fig. 2.

HOLOTYPE. Right DP⁴–M² (M51095) (Fig. 18). This specimen is chosen because it shows associated cheek teeth of one individual. Although found isolated, they were from the same sample and their matching interstitial facets demonstrate association. A left DP⁴ of identical morphology, size and wear state to the holotype DP⁴ is probably also associated (Fig. 17), but as a conservative approach it is listed as a paratype.

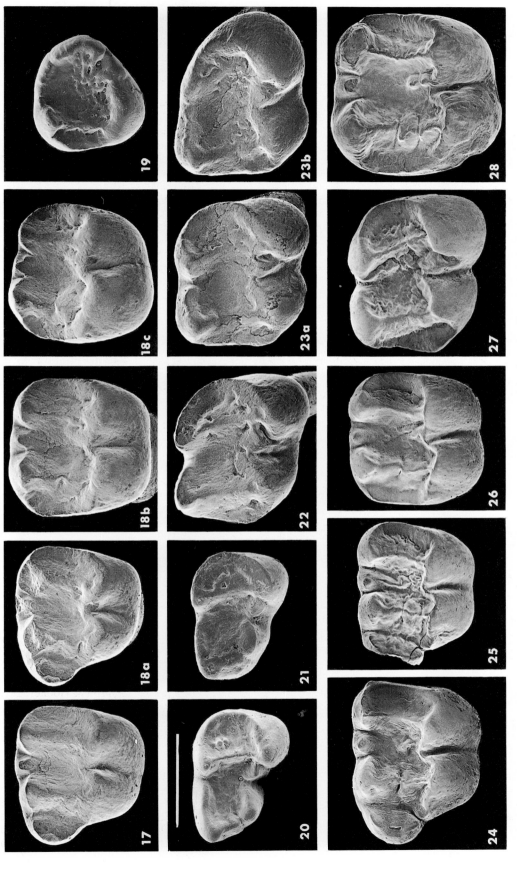

Figs 17–28 SEMs of cheek teeth in occlusal view of *Suevosciurus*, Solent Group, late Eocene, Isle of Wight. Figs 17–23, *S. bosmae* sp. nov., from green clay below How Ledge Limestone, Headon Hill. Fig. 18, **holotype**, associated right DP⁴ (a), M¹ (b), M² (c) (reversed) (M51095). Figs 17, 19–23 are paratypes: Fig. 17, left DP⁴ (M51096); Fig. 19, epoxy cast of left M³ (original GIU.941); Fig. 20, epoxy cast of right DP₄ (reversed) (original GIU.816); Fig. 21, epoxy cast of left DP₄ (original GIU.815); Fig. 22, left M₁/₂ (M51098); Fig. 23, associated right M₂ (A) and M₃ (B) (reversed) (M51097). Figs 24–27, *S. fraasi*, from the Bembridge Limestone Formation, Headon Hill. Figs 24–25, left DP⁴s (M51090–1; the latter's parastyle is slightly broken); Fig. 26, left M¹/² (M51093); Fig. 27, left M₁/₂ (M51094). Fig. 28, left M¹/² (M49489) of *S. ehingensis*, from the Bembridge Limestone Formation, Headon Hill. Scale bar = 1 mm; casts are coated with gold or gold palladium, originals are uncoated.

PARATYPES. Left DP^4 (M51096) (Fig. 17); left M^3 (GIU941) (Fig. 19); right DP$_4$ (GIU816) (Fig. 20); left DP$_4$ (GIU815) (Fig. 21); left M$_{1/2}$ (M51098) (Fig. 22); right M$_{2-3}$ (M51097) (Fig. 23).

NAME. For Dr Anneke Bosma, Utrecht, in recognition of her work on fossil rodents.

TYPE HORIZON AND LOCALITY. Green clay below the How Ledge Limestone (includes Bosma's (1974) sample locality HH2) near top of Totland Bay Member, Headon Hill Formation, SW Headon Hill, Isle of Wight.

REFERRED MATERIAL. Topotype isolated teeth M33371, M51096 and M51098–51102. Also Bosma's (1974) material in the GIU from the Totland Bay Member of Headon Hill (HH1, 2) (topotypes) and Totland Bay; and Lignite Bed in Hatherwood Limestone Member of Headon Hill (HH3, 4, B and C) (except GIU 426, 480, 492 and 499 which belong to *Treposciurus gardneri*). Additional isolated teeth from shelly lenses at base of lignite bed (same level as HH3 of Bosma 1974) (M51106–51129).

DIAGNOSIS. Mean length of M$^{1/2}$ = 1·55 mm; range 1·39–1·67 mm. P$_4^4$ absent. DP^4 mean length/width ratio 1·17–1·19. M^3 without distinct metalophule II and often with weak paraconule . DP^4 with width at paracone and protocone less than width at metacone and hypocone; no lingual parastylar notch. Lower molars nearly always with crest joining mesoconid to hypoconid.

DESCRIPTION. Much of the morphological variation has been described qualitatively by Bosma (1974). An attempt has been made in Tables 2–3 to quantify this to provide a means of comparison with *Suevosciurus authodon* from the Bartonian of Creechbarrow (Hooker 1986). Dividing lines between categories are admittedly often arbitrary and can often be documented only in relatively little-worn teeth, but despite these shortcomings this methodology can still demonstrate broad trends and differences. Only the assemblages from the green clay below the How Ledge Limestone (including HH2) near the top of the Totland Bay Member, and the shelly lenses at the base of the lignite bed in the Hatherwood Limestone (including HH3) were considered large enough for this quantitative treatment, and neither are as large as that from Creechbarrow.

Table 2 Percentage character analysis of upper cheek teeth of *Suevosciurus bosmae* from the green clay below How Ledge Limestone (HH2) and from shelly lenses at base of Lignite Bed (HH3). 'Scoring units' give states for characters described in the left-hand column. The numbers given under the tooth-headed columns on the right are percentages and refer to the number of teeth showing that particular character state. The final lines of figures in brackets in the left-hand column are the respective numbers of each tooth type represented. See Hooker (1986:296, text-fig. 31) for relevant dental nomenclature diagram.

Characters + (N) of respective teeth	Scoring units	DP^4 HH2	DP^4 HH3	M$^{1-2}$ HH2	M$^{1-2}$ HH3	M^3 HH2	M^3 HH3
Metalophule I shape: metaconule 1 only (1), ridge (2) (8) (5) (16) (8)	1	12	0	6	25		
	2	88	100	94	75		
Metalophule I joins endoloph (1), hypocone (2), metalophule II (3), endoloph and hypocone (4), hypocone and metalophule II (5), endoloph and metalophule II (6), all three (7), none (0) (8) (6) (13) (9)	0	12·5	0	7·5	22		
	1	37·5	33·2	62·0	78		
	2	37·5	16·7	23·0	0		
	3	0	16·7	0	0		
	4	0	0	0	0		
	5	0	16·7	0	0		
	6	12·5	16·7	7·5	0		
	7	0	0	0	0		
Metalophule II broken/unbroken (14) (9) (31) (22) (15) (9)	B	14	0	16	36	100	0
	U	86	100	84	64	0	0
Enamel wrinkling (0–3) (13) (9) (23) (16) (11) (10)	0	46	11	0	6	0	0
	1	23	56	56	50	9	10
	2	23	33	35	31	73	80
	3	8	0	9	13	18	10
Mesostyle size (0–4) (16) (9) (37) (23) (16) (11)	0	0	0	0	0	6	27
	1	19	0	16	22	0	0
	2	56	67	70	56	0	18
	3	19	33	14	22	44	9
	4	6	0	0	0	50	46
Mesostyle saliency: prominent (3), slight (2), non- (1), ectoflexus (0) (15) (9) (33) (20) (16) (9)	0	0	22·0	18	20	0	0
	1	20	33·5	70	55	6	33
	2	67	33·5	12	25	69	67
	3	13	11·0	0	0	25	0

Table 2 (contd)

Characters + (N) of respective teeth	Scoring units	DP^4		M$^{1-2}$		M^3	
		HH2	HH3	HH2	HH3	HH2	HH3
Mesoloph length (0–2)	0	7	0	16	33	100	100
(14) (9) (32) (21) (15) (11)	1	57	89	75	43	0	0
	2	36	11	9	24	0	0
Protolophule I broken/unbroken	B	0	0	6	11	25	89
(11) (8) (35) (19) (8) (9)	U	100	100	94	89	75	11
Metaconule 2 absence/presence	0	50	89	79	78		
(16) (9) (38) (23)	1	50	11	21	22		
Hypolophule absent (0), partial	0	85	78	77·0	75		
(1), complete (2)	1	15	22	11·5	15		
(13) (9) (35) (20)	2	0	0	11·5	10		
Posteroloph broken/unbroken lingually	B	0	11	23	17		
(7) (9) (22) (18)	U	100	89	77	83		
Mesocone absence/presence	0	100	100	100	100	87·5	100
(15) (9) (38) (21) (16) (11)	1	0	0	0	0	12·5	0
Sinus depth: shallow (1) to	1	0	0	0	0	60	45·5
deep (4)	2	12	67	10	5	40	45·5
(17) (9) (38) (21) (15) (11)	3	70	22	82	81	0	0
	4	18	11	8	14	0	9
Protostyle absence/presence	0	75	37·5	89	76	100	90
(16) (8) (38) (21) (15) (10)	1	25	62·5	11	24	0	10
Hypostyle absence/presence	0	78	100	52	50		
(9) (8) (25) (4)	1	22	0	48	50		
Paraconule absent (0), small (1),	0	0	0	0	0	0	10
large (2)	1	6	0	3	13	62	70
(16) (9) (37) (23) (13) (10)	2	94	100	97	87	38	20
Endoloph broken (1), complete (2)	1	0	11	0	12	0	18
(8) (9) (21) (17) (7) (11)	2	100	89	100	88	100	82
DP^4 parastyle bicuspid	0	44	37·5				
(9) (8)	1	56	62·5				
DP^4 mesiolingual margin concave	0	6	37·5				
(16) (8)	1	94	62·5				

Because of the small sample numbers, first and second molars were not distinguished. Many of the characters show a distribution similar to that in *S. authodon*. However, there is a general tendency in a number of the characters for one morphology in the range to dominate more than in *S. authodon*. In other words there is a slight reduction in variation. M^3 shows the greatest number of differences. Its distal reduction compared to *S. authodon* means that most specimens have lost their mesoloph and metalophule II (and, in the few that retain it, it is discontinuous), and reduced the depth of the sinus, the incidence of a protostyle, and the size of the paraconule. Mesiodistal elongation of the mesostyle is restricted to M^3, having been lost from the other upper cheek teeth. In both upper and lower cheek teeth there is an increase in enamel wrinkling intensity on DP$_4^4$–M$_2^2$. In the lower molars the hypoconulid present in some M$_{1/2}$s of *S. authodon* is not encountered in the *S. bosmae* assemblages. There is also a reduction in the incidence of the distal crest to the hypolophulid. There is a slight shift in the average position of attachment of the ectolophid to the hypolophulid in M$_3$, so that in a greater proportion the attachment is at or very close to the hypoconid. The biggest difference is in the increase in the proportion of teeth where the mesoconid is linked to the buccal side of the hypoconid by a crest, thus

Table 3 Percentage character analysis of lower cheek teeth of *Suevosciurus bosmae* from HH2 and HH3, as in Table 2.

Characters + (N) of respective teeth	Scoring units	DP$_4$ HH2	DP$_4$ HH3	M$_{1-2}$ HH2	M$_{1-2}$ HH3	M$_3$ HH2	M$_3$ HH3
Distance along hypolophulid from hypoconid of junction with ectolophid (7) (6) (20) (20) (9) (12)	$<\frac{1}{4}$	71	83	65	80	100	83
	$\frac{1}{4}$	29	0	35	20	0	17
	no link	0	17	0	0	0	0
Anteroconid size (22) (18) (9) (12)	1			27	17	0	33
	2			32	44	89	67
	3			36	39	11	0
	4			5	0	0	0
Anterolophulid absent (0), weak (1) (20) (16) (6) (12)	0			85	62·5	100	42
	1			15	37·5	0	58
Mesoconid with crest linking buccally with hypoconid (10) (6) (24) (21) (9) (12)	0	50	83	21	19	0	0
	1	50	17	79	81	100	100
Enamel wrinkling (0–3) (6) (6) (14) (16) (7) (10)	0	0	0	0	0	0	0
	1	50	100	43	44	29	20
	2	33	0	43	50	71	60
	3	17	0	14	6	0	20
Mesostylid absence/presence (10) (6) (23) (21) (9) (12)	0	100	83	96	81	44	92
	1	0	17	4	19	56	8
Ectostylid absence/presence (10) (6) (24) (21) (9) (12)	0	100	100	100	95	100	100
	1	0	0	0	5	0	0
Hypoconulid absence/presence (8) (6) (16) (17) (6) (11)	0	100	100	100	100	100	100
	1	0	0	0	0	0	0
Distal crest to hypolophulid absence/presence (8) (6) (16) (18) (6) (11)	0	100	100	100	94	100	64
	1	0	0	0	6	0	36

isolating the distal part of the sinusid as a discrete fossa. The change is most marked in M$_3$, least marked in DP$_4$. Linkage makes it impossible to allocate mesoconid length categories (cf. Hooker 1986).

None of the morphological differences between these two assemblages of *S. bosmae* appears significant. A possible exception is the M^3 protolophule I, which shows a dominantly broken or interrupted state in the Hatherwood Limestone (HH3) assemblage, in contrast to both the How Ledge Limestone (HH2) and Creechbarrow assemblages, but in common with those of later *S. fraasi* (Schmidt-Kittler 1971:42).

Discussion

Distinction of *S. bosmae* from other small *Suevosciurus*.
Hooker (1986:327) proposed that 'teeth from the Headon Beds (referred by Bosma 1974 to *S. palustris*)' should be placed in a new species, but did not name it (herein named *S. bosmae*). He explained that the assemblages in question have upper cheek teeth with a constantly larger mesostyle, whereas in all those of true *S. palustris* (admittedly few and restricted

to the type assemblage from Hoogbutsel) the mesostyle is either very small or absent. Moreover, the only two known lower molars of *S. palustris* (both IRSNB.IG18061) have a mesoconid that is not joined by a crest to the mesiobuccal corner of the hypoconid. Common occurrence of this state is shared with *S. minimus* (Schmidt-Kittler 1971: 48; pl. 2, fig. 5) and *S. authodon* (Hooker 1986: 321). Most lower cheek teeth of *S. bosmae* have the crest joining mesoconid to hypoconid, in common with assemblages of *S. fraasi* and *S. ehingensis* from southern Germany (Schmidt-Kittler 1971: 42–47). Interestingly, they are also paralleled by some individuals of *Treposciurus m. mutabilis* (Schmidt-Kittler 1971: 53, fig. 22i).

It is relatively simple to distinguish *S. bosmae* from similarly-sized *S. palustris*, but less so from other assemblages of small *Suevosciurus*. Hooker (1986: 325), using mainly published measurements, combined data from the Hampshire Basin and Bavaria (southern Germany) to produce a phylogenetic pattern of change in *Suevosciurus* assemblages through time. It essentially followed the concept of Schmidt-Kittler (1971) of two evolving lineages in the latest Eocene and Oligocene of southern Germany, except that it removed *S.*

minimus from a common ancestral position, replacing it with the taxon here named *S. bosmae*. An unsolved problem over the two lineage model of Schmidt-Kittler was the near total reliance on size for distinguishing each lineage, which itself undergoes size increase with time. On this basis, therefore, one cannot identify certain of the assemblages without recourse to knowledge of their age. Bosma (1974: 41–43) thus resorted to an arbitrary division of species on size, which could be considered more parsimonious in the absence of other morphological evidence. Hooker (1986: 326–327) found a gradual increase in length/width proportions of DP^4, through at least the lower part of the sequence, which served to distinguish further some of the similarly-sized assemblages, although problems remained in the probable region of differentiation of *S. fraasi* and *S. ehingensis*.

One of the assemblages which Schmidt-Kittler (1971) placed in *S. fraasi* and Bosma (1974) placed in *S. palustris* is from Ehrenstein 1. This fissure filling contains faunas of two different ages, labelled A and B, so intermixed that they can only be distinguished by comparing each element with those in stratified deposits of known age (Schmidt-Kittler 1969, 1971). The later fauna (B) dates from just after the Grande Coupure, the earlier (A) from the middle of the late Eocene, approximately the age of the Lacey's Farm Limestone Member, Headon Hill Formation of the Isle of Wight. Schmidt-Kittler (1971) considered that of the three *Suevosciurus* species, *S. minimus*, the most primitive, came from the A fauna, whereas *S. fraasi* and *S. ehingensis* came from the B fauna. This means that from just pre-Grande Coupure time (Bernloch 1A/Weissenberg 2) onwards, the otherwise constantly sized *S. fraasi* lineage underwent a rapid size decrease (Ehrenstein 1B) followed by similar increase soon afterwards (Ehingen 12) (see Hooker 1986: text-fig. 38). Schmidt-Kittler (1971: 47) noted some minor morphological differences between his *S. fraasi* and *S. ehingensis* lineages: slightly blunter and more voluminous main cusps and somewhat weaker parastyle on DP^4 in the latter. In fact there is a tendency for DP^4s of post-Grande Coupure assemblages of *S. fraasi* to be dominated by the two morphs that Schmidt-Kittler (1971: 42) described: lingually displaced parastyle, causing a very oblique buccal parastyle margin (his fig. 16b); and notch in the mesial outline just lingual to the parastyle (his fig. 16d). In *S. ehingensis* these morphs are either rare or less clearly developed. The pre-Grande Coupure assemblages referred to *S. fraasi* are less distinct but in common with the post-Grande Coupure ones tend to have the widths of DP^4 across paracone–protocone and metacone–hypocone approximately equal. In *S. ehingensis* the mesial width tends to be slightly shorter than the distal width, in common with *S. bosmae* and *S. authodon* (Fig. 30). DP^4s referred to *S. fraasi* in the Ehrenstein 1 assemblage are morphologically less distinct than the other post-Grande Coupure assemblages and, together with their small size, would thus fit better in the Ehrenstein 1A than the 1B fauna.

Recently, Heissig (1987) has described a new small species of *Suevosciurus*, *S. dehmi*, from the immediately post-Grande Coupure Bavarian fissure filling of Mohren 31. The type assemblage is slightly larger than *S. bosmae* but overlaps slightly with it. It could weaken the evidence for the timing of the speciation event envisaged here by potentially supporting the B age for Ehrenstein 1 *S. fraasi* (the intermediate-sized species), through the latter's possible identification as *S. dehmi*. It does not, however, disprove it. Heissig (1987:102, fig. 1) did not include the plot of the intermediate-sized

Suevosciurus from Ehrenstein 1, but, from Schmidt-Kittler's (1971) text-fig. 20, it would superimpose the type assemblage of *S. dehmi*. Unfortunately, Heissig did not diagnose *S. dehmi* on characters other than size and it is thus difficult to fit it into a scheme based on morphology. However, his figure (Heissig 1987: pl. 1) of the holotype right dentary with DP_4 (not P_4), M_2 and M_3 shows the molar mesoconids joining the hypoconids as is usual for *S. fraasi*, *S. ehingensis* and *S. bosmae*. If the intermediate-sized *Suevosciurus* from Ehrenstein 1 is indeed *S. dehmi*, then its relationships appear closer to *S. fraasi* than to *S. ehingensis*, perhaps introducing paraphyly for *S. fraasi*. However, decision must await publication of morphological details of *S. dehmi*.

Evidence for a late Eocene speciation event.
To clarify ideas of relationships of the advanced species of *Suevosciurus* (i.e. those that have lost P_4^4), a cladistic analysis is presented here, using the admittedly variable morphological differences in addition to size. *S. authodon* is used as outgroup to polarize the characters. The placement of *S. palustris* is doubtful, as neither DP_4^4 nor P_4^4 tooth types are known, and assumes that characters 7 and 8 are in the advanced state (Fig. 29).

If the Ehrenstein 1 assemblage referred to *S. fraasi* is from the A fauna, as advocated above, it is envisaged that initial differentiation of *S. fraasi* and *S. ehingensis* from the probable ancestral species *S. bosmae* took place, in the former by protocone expansion causing mesial broadening with incipient lingual parastyle notching of DP^4 but with little size increase, and in the latter simply by a greater increase in size unaccompanied by DP^4 shape changes (Figs 30, 31). According to this model the most primitive *S. fraasi* assemblage would be that of Ehrenstein 1, whereas the most primitive *S. ehingensis* assemblage would be that of Lacey's Farm Quarry (Lacey's Farm Limestone Member) (identified on size as *S. fraasi* by Bosma & Insole 1976). There seems also to be a slight size increase of DP^4 over the molars in *S. fraasi*, so that DP^4s from Ehrenstein 1 are about the same size as those from Lacey's Farm Quarry, whereas the molars of the former are smaller; by the time of Weissenburg 8, *S. ehingensis* DP^4s had enlarged proportionally also (Fig. 30). Subsequent patterns of change, in addition to size increase in both lineages (causing advanced assemblages of *S. fraasi* to have character 3), involve repeated elongations and shortenings of DP^4 (Hooker 1986: text-fig. 39), as well as subtle shape changes. None of these, however, affects recognition of the diagnostic *S. fraasi* DP^4s provided assemblages are large enough. These changes thus comprise variation within a lineage which could be discriminated taxonomically at the level of stratigraphical subspecies (e.g. as Franzen, 1968, has done for *Palaeotherium*).

In further support of the speciation model, rare specimens of both *S. fraasi* and *S. ehingensis* have been found in the argillaceous beds of the Bembridge Limestone Formation of Headon Hill (= HH6–7 of Bosma, 1974). *S. ehingensis* is represented by three $M^{1/2}$s slightly larger than those from Lacey's Farm Quarry, whilst *S. fraasi* is represented by three DP^4s, one $M^{1/2}$ and one $M_{1/2}$ (Figs 24–28). The teeth of *S. fraasi* are all significantly smaller than the equivalent tooth types from Lacey's Farm Quarry; moreover, the DP^4s have a greater length/width ratio, equal widths at both paracone–protocone and metacone–hypocone, and larger, more prominent parastyle with lingual notch.

The highest definite record of *S. bosmae* is from HH4. A few teeth from higher up in marly beds at the top of the

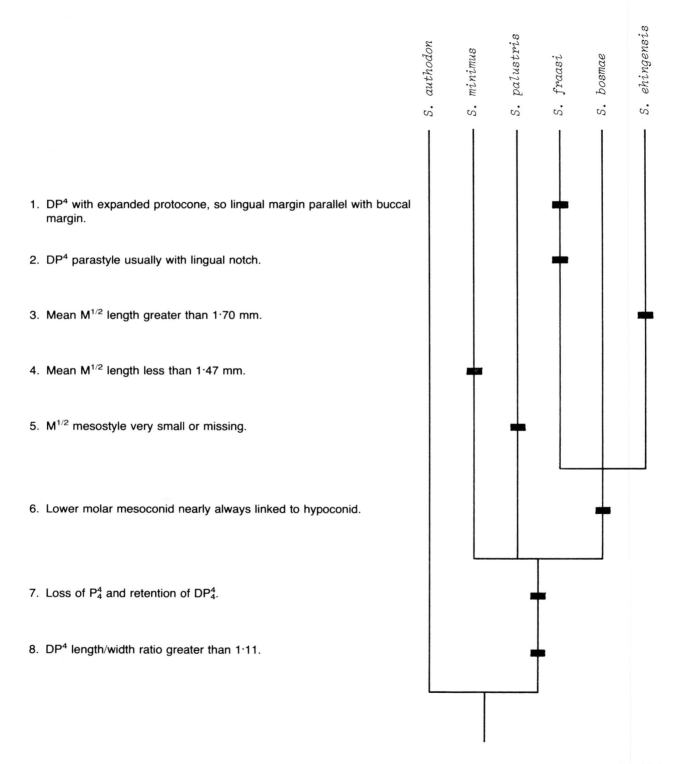

1. DP⁴ with expanded protocone, so lingual margin parallel with buccal margin.

2. DP⁴ parastyle usually with lingual notch.

3. Mean M^{1/2} length greater than 1·70 mm.

4. Mean M^{1/2} length less than 1·47 mm.

5. M^{1/2} mesostyle very small or missing.

6. Lower molar mesoconid nearly always linked to hypoconid.

7. Loss of P_4^4 and retention of DP_4^4.

8. DP⁴ length/width ratio greater than 1·11.

Fig. 29 Cladogram of species of *Suevosciurus*. Synapomorphies and autapomorphies are indicated by broad bar. See Fig. 16 for synapomorphies of the genus.

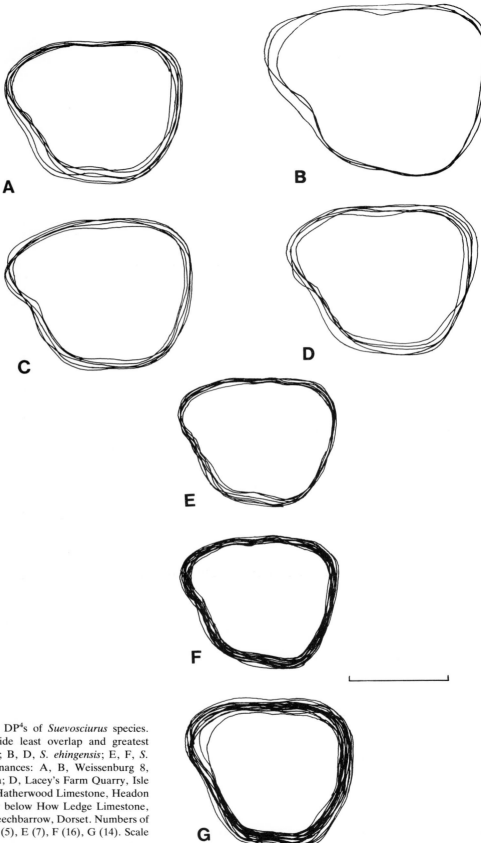

Fig. 30 Superimposed outlines of DP⁴s of *Suevosciurus* species. Superimposed manually to provide least overlap and greatest shape alignment. A, C, *S. fraasi*; B, D, *S. ehingensis*; E, F, *S. bosmae*; G, *S. authodon*. Provenances: A, B, Weissenburg 8, Bavaria; C, Ehrenstein 1, Bavaria; D, Lacey's Farm Quarry, Isle of Wight; E, base of lignite bed, Hatherwood Limestone, Headon Hill, Isle of Wight; F, green clay below How Ledge Limestone, Headon Hill, Isle of Wight; G, Creechbarrow, Dorset. Numbers of specimens: A (6), B (3), C (5), D (5), E (7), F (16), G (14). Scale bar = 1 mm.

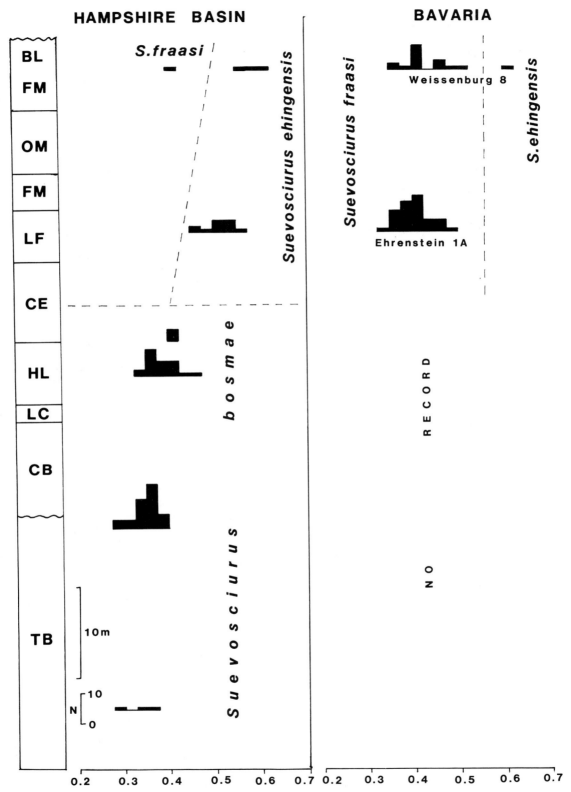

Fig. 31 Histograms in stratigraphic order of $\log_{10}$ (length) × (width) of $M^{1/2}$ of *Suevosciurus* species, from the late Eocene Solent Group sequence in the central Hampshire Basin and relevant fissure fillings of equivalent age in Bavaria, southern Germany. N = number of specimens; m = metres of section; BL FM = Bembridge Limestone Formation; OM = Osborne Marls Member; FM = Fishbourne Member; LF = Lacey's Farm Limestone Member; CE = Cliff End Member; HL = Hatherwood Limestone Member; LC = Linstone Chine Member; CB = Colwell Bay Member; TB = Totland Bay Member. The lower histogram in TB refers to the Hordle Mammal Bed.

Hatherwood Limestone Member of slightly larger mean size (Fig. 31) may represent *S. bosmae*, but the sample as yet lacks the distinctive DP^4. The next record of a *Suevosciurus* is of *S. ehingensis* from the Lacey's Farm Limestone Member, penecontemporaneous with the earliest record of *S. fraasi* at Ehrenstein 1. The evidence available suggests that the speciation event took place within the time represented by deposition of the intervening Cliff End Member at Headon Hill and may have resulted from isolation of populations in the southern English and Bavarian areas respectively. At some time after the deposition of the Lacey's Farm Limestone, and before deposition of the Bembridge Limestone, renewed dispersal to both areas became possible. Species distribution patterns in other genera of mammals from this sequence do not point to other contemporaneous speciation events, although from at least this time until well into the Oligocene, Bavaria had largely endemic faunas (Schmidt-Kittler & Vianey-Liaud 1975; Heissig 1978), and the earlier northern European region as portrayed by Franzen (1968) was split into two by Schmidt-Kittler & Vianey-Liaud (1975), with the boundary at the Rhine Graben. Ziegler (1982) records the Rhine Graben as a site of vulcanism and marine transgression in the late Eocene, which may well have been the isolating mechanism for the *Suevosciurus* speciation. If so, however, it did not prevent the newly formed species from migrating subsequently in both directions.

Speciation patterns

Cladogenetic speciation events have been recorded for several groups of mammals in the dense early Eocene record of the Bighorn Basin, Wyoming, U.S.A., in addition to the more obvious anagenetic events (e.g. Gingerich 1974, 1976, 1977, 1980; Gingerich & Simons 1977). In most cases, however, as one traces two lineages back in time, it is possible to follow only one of them right to an ancestral species, although the pattern of change implies derivation of both from a common ancestor (e.g. Gingerich 1976). A similar pattern has also been described for European Eocene primates (Godinot 1985). At first described as parapatric, this type of cladogenetic speciation was later considered to be allopatric, but where most of the morphological differentiation took place subsequently anagenetically and sympatrically (Gingerich 1977: 491–493). An important problem, however, remains: the lack of similarly dense fossil sequences in areas outside the Bighorn Basin, where the missing branch segment might be represented. Although the late Eocene/early Oligocene mammalian record in Europe is not as dense as in the early Eocene of the Bighorn Basin, and although the sequence in Bavaria is based on biostratigraphy of other mammals, not on superposition, a roughly equal resolution of mammalian faunal succession is recognizable in two European areas. Moreover, in each, it is possible to trace the two lineages of *S. fraasi* and *S. ehingensis* back in time until the former appears in southern England and the latter appears in southern Germany. Earlier than this in the late Eocene of southern England, only a single species (*S. bosmae*) occurs, which is primitive with respect to both *S. fraasi* and *S. ehingensis*. The pattern envisaged here is consistent with a more traditional idea of allopatric speciation, where geographic isolation results in morphologic as well as genetic differentiation before remixing of populations (the 'dumbbell' model – White 1978; Mayr 1982). It may simply have happened that isolation time lasted longer here. Alternatively, smaller population sizes in Eocene 'island' Europe

may have produced more rapid character changes. For instance, geographical ranges of some of the critical Bighorn Basin mammals seem to have been large, extending at least from Wyoming to New Mexico (e.g. Gingerich & Simons 1977). The distance between the two European areas under consideration is about 900 km and the maximum area potentially involved (i.e. delimited to the south by a line drawn from Paris to Geneva, as *Suevosciurus* only occurs very rarely further south), judged from palaeogeographic reconstructions (e.g. compilations by Ziegler 1982 and Hooker 1986) is about 300,000 km^2. This is no bigger than the state of Wyoming alone.

It is interesting to note that the newly discovered specimens of *S. fraasi* and *S. ehingensis* in the Bembridge Limestone tend to occur at different levels. Moreover, the former is associated with other taxa, such as primates, an apatemyid and a bat, all small forms which suggest a forested environment. In contrast, the latter is associated with other taxa such as the ungulates *Plagiolophus* and *Diplobune*, which suggest a slightly more open environment. This may indicate habitat differences which were directly related to their speciation. Taphonomic study of this sequence in progress may shed further light on this matter.

There are various potential tests for this speciation event. For instance, to find an Ehrenstein 1 stage *S. fraasi* in a Bavarian fissure containing a single-aged fauna would determine which of the two possible ages was right. The undescribed material from Ehrenstein 2, 3 and 6, Herrlingen 3 and Arnegg 3, listed as '*Suevosciurus minimus – fraasi* (Übergangsform)' by Schmidt-Kittler (1977) could provide the answer. Moreover, the finding of even earlier fissures in this area containing *S. bosmae* would support the isolation model. Conversely, the finding of *S. fraasi* in the Lacey's Farm Limestone Member or contemporaneous strata in the Isle of Wight, or of *S. ehingensis* in a fissure filling the same age as the Ehrenstein 1A fauna in Bavaria, would suggest that the speciation pattern was instead like that documented in the Bighorn Basin for, e.g., *Hyopsodus* and *Cantius* (Gingerich 1977).

ACKNOWLEDGEMENTS. I would like to thank Mr R. G. Gardner for presenting the type series of *T. gardneri* and other specimens to the Natural History Museum; Mr & Mrs J. Smith and the National Trust (in particular Messrs J. Simson, E. Brown and A. Tutton) for access to important localities. The following museum curators kindly provided access to collections in their care: Professor R. J. G. Savage and Dr L. Loeffler (Bristol); Dr M. Weidmann (Lausanne); Dr H. de Bruijn (GIU); Dr P. Sartenaer (IRSNB); Dr J.-L. Hartenberger (Montpellier); and Professors V. Fahlbusch and R. Dehm (BSPG). Mr N. V. Hayes of the Photographic Studio, Natural History Museum printed the photographs.

REFERENCES

Bosma, A. A. 1974. Rodent biostratigraphy of the Eocene–Oligocene transitional strata of the Isle of Wight. *Utrecht Micropaleontological Bulletin. Special Publications*, **1**. 128 pp., 7 pls, 36 figs.

—— & **Insole, A. N.** 1976. Pseudosciuridae (Rodentia, Mammalia) from the Osborne Beds (Headonian), Isle of Wight, England. *Proceedings Koninklijke Nederlandse Akademie van Wetenschappen*, Amsterdam, (B) **79**: 1–8, 1 pl.

Bristow, H. W., Reid, C. & Strahan, A. 1889. *The Geology of the Isle of Wight* (2nd edn). xiv + 349 pp. London, Mem. Geol. Surv. U.K.

Collinson, M. E. & Hooker, J. J. 1987. Vegetational and mammalian faunal changes in the Early Tertiary of southern England. *In* Friis, E. M., Chaloner,

W. G. & Crane, P. R. (eds), *The Origins of Angiosperms and their Biological consequences*: 259–304. Cambridge.

Cray, P. E. 1973. Marsupialia, Insectivora, Primates, Creodonta and Carnivora from the Headon Beds (Upper Eocene) of southern England. *Bulletin of the British Museum (Natural History)*, London, (Geol.) **23** (1): 1–102.

Dehm, R. 1937. Über die alttertiäre Nager Familie Pseudosciuridae und ihr Entwicklung. *Neues Jahrbuch für Mineralogie, Geologie und Paläontologie*. Stuttgart, Beilagebände (B) **77**: 268–290.

Franzen, J.-L. 1968. *Revision der Gattung Palaeotherium Cuvier 1804 (Palaeotheriidae, Perissodactyla, Mammalia)*. **1**, 181 pp., 20 figs; **2**, 35 pls, 15 tabs. Freiburg, Inaugural-Dissertation zur Erlangung der Doktorwurde der Naturwissenschaftlich-mathematischen Facultät der Albert-Ludwigs-Universität zu Freiburg i. Br.

Gingerich, P. D. 1974. Stratigraphic record of Early Eocene *Hyopsodus* and the geometry of mammalian phylogeny. *Nature, London*, **248** (5444): 107–109.

—— 1976. Paleontology and phylogeny: patterns of evolution at the species level in early Tertiary mammals. *American Journal of Science*, New Haven, Conn., **276**: 1–28.

—— 1977. Patterns of evolution in the mammalian fossil record. *In* Hallam, A. (ed.), *Patterns of Evolution*: 469–500. Amsterdam.

—— 1980. Evolutionary patterns in early Cenozoic mammals. *Annual Review of Earth and Planetary Sciences*, Palo Alto, Calif., **8**: 407–424.

—— & Simons, E. L. 1977. Systematics, phylogeny, and evolution of early Eocene Adapidae (Mammalia, Primates) in North America. *Contributions from the Museum of Paleontology, University of Michigan*, Ann Arbor, **24** (22): 245–279.

Godinot, M. 1985. Evolutionary implications of morphological changes in Palaeogene primates. *In* Cope, J. C. W. & Skelton, P. W. (eds), Evolutionary case histories from the fossil record. *Special Papers in Palaeontology*, London, **33**: 39–47.

Hartenberger, J.-L. 1973. Étude systématique des Theridomyoidea (Rodentia) de l'Eocène supérieur. *Mémoires de la Société Géologique de la France*, Paris, (NS) **52** (Mém. 117): 1–76.

Heissig, K. 1978. Fossilführende Spaltenfüllungen Süddeutschlands und die Ökologie ihrer oligozänen Huftiere. *Mitteilungen der Bayerischen Staatssammlung für Paläontologie und historische Geologie*, München, **18**: 237–288.

—— 1987. Changes in the rodent and ungulate fauna in the Oligocene fissure fillings of Germany. *Münchner Geowissenschaftliche Abhandlungen*, (A) **10**: 101–108.

Hooker, J. J. 1986. Mammals from the Bartonian (middle/late Eocene) of the Hampshire Basin, southern England. *Bulletin of the British Museum (Natural History)*, London, (Geol.) **39** (4): 191–478.

—— 1987. Mammalian faunal events in the English Hampshire Basin (late Eocene – early Oligocene) and their application to European biostratigraphy. *Münchner Geowissenschaftliche Abhandlungen*, (A) **10**: 109–115.

—— 1989. British mammals in the Tertiary Period. *Biological Journal of the Linnean Society*, London, **38**: 9–21.

—— & Insole, A. N. 1980. The distribution of mammals in the English Palaeogene. *In* Hooker, J. J., Insole, A. N., Moody, R. T. J., Walker, C. A. & Ward, D. J., The distribution of cartilaginous fish, turtles, birds and mammals in the British Palaeogene. *Tertiary Research*, Rotterdam, **3** (1): 1–45.

Insole, A. N. & Daley, B. 1985. A revision of the lithostratigraphical nomenclature of the Late Eocene and Early Oligocene strata of the Hampshire Basin, southern England. *Tertiary Research*, Leiden, **7** (3): 67–100.

Major, C. I. F. 1873. Nagerüberreste aus Bohnerzen Süddeutschlands und der Schweiz. *Palaeontographica*, Cassel, **22** (2): 75–130, taf. III–VI (initials cited as 'C. J. F.').

Matthews, S. C. 1973. Notes on open nomenclature and on synonymy lists. *Palaeontology*, London, **16** (4): 713–719.

Mayr, E. 1982. Speciation and macroevolution. *Evolution*, Lancaster, **36** (6): 1119–1132.

Misonne, X. 1957. Mammifères oligocènes de Hoogbutsel et Hoeleden; 1 – Rongeurs et ongules. *Bulletin de l'Institut Royal des Sciences Naturelles de Belgique*, Bruxelles, **33** (51): 1–61.

Russell, D. E. *et al.* 1982. Tetrapods of the Northwest European Tertiary Basin. *Geologisches Jahrbuch*, Hannover, (A) **60**: 1–74.

Schlosser, M. 1884. Die Nager der europäischen Tertiärs. *Palaeontographica*, Cassel, **31** (3): 1–143 (19–161), pls 1–8 (5–12).

Schmidt-Kittler, N. 1969. Eine alttertiäre Spaltenfüllung von Ehrenstein, westlich Ulm. *Mitteilungen der Bayerischen Staatssammlung für Paläontologie und historische Geologie*, München, **9**: 201–208.

—— 1970. Ein neue Pseudosciuride von Ehrenstein, westlich Ulm. *Mitteilungen der Bayerischen Staatssammlung für Paläontologie und historische Geologie*, München, **10**: 433–440.

—— 1971. Odontologische Untersuchungen an Pseudosciuriden (Rodentia, Mammalia) des Alttertiärs. *Abhandlungen der Bayerischen Akademie der Wissenschaften*, München, Math.-naturw. Kl., (NF) **150**: 1–133.

—— 1977. Neue Primatenfunde aus unteroligozänen Karstspaltenfüllungen Süddeutschlands. *Mitteilungen der Bayerischen Staatssammlung für Paläontologie und historische Geologie*, München, **17**: 177–195.

—— & Vianey-Liaud, M. 1975. Les relations entre les faunes de rongeurs d'Allemagne du Sud de France pendant l'Oligocène. *Compte Rendu Hebdomadaire des Séances de l'Académie des Sciences*, Paris, (D) **281** (9): 511–514.

Sudre, J. 1978. Les artiodactyles de l'Eocène moyen et supérieur d'Europe occidentale (systématique et évolution). *Mémoires et Travaux de l'Institut de Montpellier*, **7**: 1–229.

Thaler, L. 1966. Les rongeurs fossiles du Bas-Languedoc dans leurs rapports avec l'histoire des faunes et la stratigraphie du Tertiaire d'Europe. *Mémoires du Muséum National d'Histoire Naturelle de Paris*, **17**: 1–296.

White, M. J. D. 1978. *Modes of Speciation*. 455 pp. San Francisco.

Ziegler, P. A. 1982. *Geological Atlas of Western and Central Europe*. **1** (text), 130 pp.; **2** (atlas), 40 enclosures. The Hague (Shell Int. Petrol. Maat. B.V.).

Bull. Br. Mus. nat. Hist. (Geol.) **47** (1): 51–100

Issued 29 August 1991

Upper Palaeozoic Anomalodesmatan Bivalvia

N. J. MORRIS

Department of Palaeontology, British Museum (Natural History), Cromwell Road, London SW7 5BD

J. M. DICKINS

Bureau of Mineral Resources, Canberra, Australia

K. ASTAFIEVA-URBAITIS

Muzei Zemlevedeniia, Moscow State University, Moscow 119899, USSR

CONTENTS

SYNOPSIS. The systematics of the Late Palaeozoic Anomalodesmata (Bivalvia) are considered. Two families, the Sanguinolitidae and Permophoridae, are recognized which are considered to include the ancestors of most of the post-Palaeozoic taxa. They are united by the presence of periostracal spicules in many species of both families, which we regard as a synapomorphy. The new genus *Gilbertsonia* is recognized in the Sanguinolitidae and the two new genera *Siliquimya* and *Bowlandia* are described in the Permophoridae. The Grammysiidae are considered to be a paraphylum which includes the stem group of the Sanguinolitidae. The new Subfamily Cuneamyinae is recognized within the Grammysiidae. We retain three other family group taxa in the Anomalodesmata, the Orthonotidae, the Solenomorphidae (tentatively including a new Subfamily Promacrinae) and the Edmondiacea, which are better placed there than elsewhere, but we have difficulty finding reliable synapomorphies linking them to mainstream forms. Of these three families, we only have evidence that the Solenomorphidae survived beyond the Palaeozoic. Various Upper Palaeozoic Anomalodesmata demonstrate development of a posterior gape, a deep pallial sinus or elongation of the shell, which (by analogy with living taxa) indicate the development of deep sessile burrowing, while other taxa are interpreted as shallow infaunal, slightly mobile burrowers or infaunal nestlers. Some we interpret as crevice dwellers, cavicolous or even epifaunal. The Anomalodesmata reached a high degree of species diversity in the Upper Palaeozoic, with the result that among infaunal bivalves they were commonly the most species numerous bivalve subclass. This species diversity has subsequently increased only gradually, at a much slower rate than other infaunal bivalves. They are now commonly outnumbered by the Lucinacea, Mactracea, Veneracea and Tellinacea in many shallow marine habitats. All Upper Palaeozoic Anomalodesmata have a parivincular, opisthodetic ligament mounted on nymphs which, in conjunction with elongation of the animals posterior to the umbones, leaves the distal part of the dorsal shell margin joined only by periostracum, best interpreted as primitively present rather than the result of secondary fusion. The hinge systems have few or no hinge teeth. No Palaeozoic anomalodesmatan has as yet been discovered with an internal ligament, typical of a number of Mesozoic and surviving lineages.

INTRODUCTION

During the last two decades understanding of the morphology and ecology of Upper Palaeozoic bivalves (pelecypods) has developed rapidly and, in particular, the understanding of the adaptation of form to burrowing habits and the ancestral relationship of Mesozoic forms. An early analysis by Newell (1956) was followed by the work of Runnegar (1965, 1966, 1967, 1968, 1974), Waterhouse (1965, 1966, 1969*a*, 1969*b*) and Runnegar & Newell (1974). Astafieva-Urbaitis has extensively examined the morphology and relationships of Carboniferous forms (1962, 1964, 1970, 1973, 1974*a*, 1974*b*), and Muromzeva has described both Carboniferous and Permian taxa from the Soviet Arctic (1974, 1984).

The present paper examines the morphology and relation-ships of forms which have been referred to *Sanguinolites* M'Coy 1844, and *Allorisma* King 1844, and their relationships to each other and to younger forms are considered. (Hind (1900: 311) referred these genera to the Family Coelonotidae M'Coy 1855. The family name Caelonotidae (not Coelono-tidae) was used by M'Coy (1852: 275). The name Caelonotidae is apparently unavailable as it is not based on a valid generic name (I.C.Z.N. Art. 11 (e)), as observed by Runnegar (1967: 27)). Our work should be considered within the overall framework presented by Runnegar & Newell (1974) and Runnegar (1974). One group of species, which for the present is referred to *Pleurophorella* Girty (1904), is shown to be readily separable and should be placed in the Family Permophoridae. Hitherto, this family has been placed in the Superfamily Carditacea; in this paper its position there is regarded as unlikely as is also its relation-ship with the heterodonts. Other forms are placed in the

Edmondiidae and Sanguinolitidae and their family relationships are discussed.

Astafieva-Urbaitis (1974a) has investigated the relationships of species placed in *Sanguinolites*. She concluded that a number of distinct groups have been included in the genus *Sanguinolites*. We agree with her conclusions, and we make further proposals to help in resolving this problem, based on the examination of the extensive collections in the British Museum (Natural History), London, including Hind's material. Astafieva-Urbaitis (1974b) also discussed the characters and relationships of *Praeundulomya*, which she concluded was related to *Wilkingia*, and she placed it within the Family Sanguinolitidae. In 1973 she had proposed the Subfamily Undulomyinae (of the Sanguinolitidae) for *Wilkingia*, *Praeundulomya*, and *Undulomya*. In 1983 with Dickins she introduced a new name *Dulunomya* for species which they regarded as morphologically intermediate between *Wilkingia* and *Praeundulomya*. Here we reassess these genera based on their type species and include *Exochorhynchus*. We consider a taxon in this subfamily to be ancestral to *Pholadomya*.

SHELL STRUCTURE

In common with the living Anomalodesmata, the shell thickness of the Sanguinolitidae is variable: it is thin in *Sanguinolites* and *Wilkingia* and rather thicker in *Pleurophorella*. Most species of these genera have a pustulose surface. In the more spectacular forms the pustules are calcareous spikes (Aller 1974), which occur in *Wilkingia*, *Praeundulomya*, *Pholadella*, *Cimitaria* and *Chaenomya* where they are arranged in prominent radial rows. In *Pleurophorella striatogranulatus* they are better described as pustules.

Periostracal calcareous structures occur in several groups of bivalves (Carter & Aller 1975) but are not clearly present in all the taxa we consider to belong to the Anomalodesmata; this may sometimes be due to lack of preservation or sometimes their real absence. Where they do occur, however, in spike-like form, we consider that they indicate that the taxa possessing them do belong to the subclass. We consider the absence of periostracal calcareous structures in certain Anomalodesmata to be a character of taxonomic value at a level lower than subclass.

Periostracal calcareous structures are very uncommonly preserved in the Edmondiacea. Their presence in one species of the Family Megadesmidae (Runnegar 1965) and in Russian specimens of the genus *Allorisma* (Astafieva-Urbaitis & Dickins, personal observation) leads us to accept the inclusion of the superfamily within the Anomalodesmata. Other records of calcareous surface spicules that we have checked in the Edmondiacea have proved to be spurious. Pustules are also unknown in the Orthonotidae, a fact which supports the view that they should not be included within the Pholadomyoida.

Spicules are clearly preserved in *Pachymyonia* cf. *occidentalis* Dickins (1963: pl. 5, fig. 20) from the Permian Fossil Cliff Formation of Irwin River, West Australia, but this species may be related to *Sanguinolites argutus* Phillips, from the Viséan of Britain, rather than the Megadesmidae. Wilson (1960: 111) recorded fine, close-set striae radiating from the umbones of *Edmondia sulcata* (i.e. *Allorisma sulcata* of this paper). He considered these to be internal representatives of the rows of minute tubercles on the exterior of the shell

illustrated by Hind (1899: pl. 35, fig. 11a). Hind's piece of shell, however, is apparently a *Wilkingia* and the radiating striae seem to be concerned with the attachment of the mantle to the shell and not connected with external pustules.

We have been able to study the shell structure in just one species, *Myofossa costellata* (M'Coy, 1851a) where an internal nacreous layer is perfectly preserved (Fig. 1). A very thin

Fig. 1 Nacreous shell structure of the inner ostracum of *Myofossa costellata* (M'Coy); BM L46425, oblique stereoscan view of broken shell, ×720; see also Fig. 12d.

outer layer is less well preserved and is in a blocky, recrystallized form, but we have been able to interpret it by its general appearance as a thin prismatic outer layer that was originally aragonite. A simple thin myostracal layer of short aragonite prisms was identified, as in other Mollusca.

We suspect that the calcareous shell of the Sanguinolitidae consists of an outer layer of aragonite prisms, usually with a pustulose surface, and an inner layer of nacreous structure subdivided by a sheet of myostracum. This combination of aragonite prisms with an inner nacreous layer is generally considered primitive for the Mollusca (Taylor 1973), and we believe it to be the primitive condition of the Anomalodesmata. We consider that non-nacreoprismatic shells have evolved on several separate occasions in the descendants of the Palaeozoic Anomalodesmata. Intermediate stages are recognized in the Myopholadidae which we interpret as early representatives of the Pholadidacea (Taylor, Kennedy & Hall 1973). The nacreous layer may have given rise to homogeneous structure in the Ceratomyidae, although this structure could already have been present in the Edmondiidae and the Solenomorphidae, one species of which may have been their ancestor. This may also have happened in the Pleuromyidae and their descendants such as *Panopea*, and in *Gastrochaena* and *Myoconcha*, It may also have evolved later into cross lamellar structure in the Myopholadidae and Myidae.

In the Corbulacea, crossed and complex crossed-lamellar structure may also have evolved from the primitive anomalodesmatid shell, but an alternative possible derivation of this group is from the heteroconchs *via* the Myophoricardiidae. *Myopholas* and some species of *Panopea* retain the outer layer of aragonite prisms complete with pustules, at least in their early growth stages, but these are lost in the later Pholadacea, the Myidae and *Hiatella*.

HINGES

We have examined well-preserved hinges in several species of *Sanguinolites, Pleurophorella, Myofossa, Wilkingia* and *Chaenomya* and have compared them with the better known hinges of the Edmondiidae, Permophoridae, Megadesmidae and the post-Palaeozoic Anomalodesmata. The sanguinolitid dorsal margin is usually straight and long, the valves being in juxtaposition from just in front of the umbones to a point above or just posterior to the posterior adductor scar. Nymphs are present, but are normally not well developed; their anterior point is immediately below the umbones, and posterior to this point (Figs 6, 9, 12, 13, 32) they vary from short to long.

We have observed ligament material only in *Pleurophorella* sp., *Sanguinolites costatus, Chaenomya leavenworthensis* and *Edmondia* sp. Ligaments have previously been described in the Megadesmidae (Runnegar, 1968). In that family the ligament does not extend posteriorly very far from the nymph. Runnegar recognized a small extension to the ligament beyond the partly fibrous 'C spring' ligament mounted on the nymphs. In his text-figure 1a he calls this the 'fusion(?) layer', following the interpretation of modern ligaments by Yonge (1957), Owen, Trueman & Yonge (1953) and others. In the Edmondiidae (Waterhouse 1966, Runnegar & Newell 1974) the ligament nymph is relatively more elongate.

The Upper Palaeozoic taxa here included in the Pholadomyacea have a limited range of ligament types. All have structures which are properly described as nymphs; a number of species have the remains of ligament attached to these nymphs. In what we interpret as the more primitive forms, the nymph is set in a clearly defined escutcheon, usually set between dorsal carinae, which extends from the initial growth point between the umbones to the posterior dorsal corner of the shell. In most of these taxa the dorsal margins within the escutcheon are straight and come into close contact with each

other, whether the valves are preserved in the live open or closed position. In such forms we consider that the periostracal covering of the ligament continued beyond the distal end of the nymphs, joining the shells as periostracal ligament. This type of ligament has been termed 'fusion layer' by Yonge (1957). However, in the case of these Upper Palaeozoic anomalodesmatans we consider the term to be inappropriate. We consider that this periostracum was not formed by fused mantle edges but was more likely to have been laid down by the original, primitive dorsal skin of the animal. It may have been deposited initially by the inner surface of the outer mantle fold at the posterior of the shell and then thickened by the outer surface of the skin along the dorsum. The dorsum and the periostracal ligament may have been extended from a more primitive shorter state by posterior hypertrophy of these elongate infaunal animals.

Modifications of the primitive ligament pattern that occur in the Upper Palaeozoic Anomalodesmata include the presence of nymphs where the lamellar and fibrous ligament layers are attached. In a hypothetical, simplest form, all the layers of the ligament would extend approximately the same distance distally along the dorsum. This arrangement occurs today only in the Mytilacea, where it is modified by the peculiar nature of the calcareous material joining the ligament to the shell. In all the Pholadomyacea we have studied, the distinct nymph does not run to the end of the escutcheon. The nymph is usually relatively short in the taxa we interpret as having been entirely infaunal; this condition occurs in *Grammysioidea, Sedgwickia*, and *Chaenomya*. It is slightly longer in the Undulomyinae, longer still in *Sanguinolites*, and quite long in many of the Permophoridae, which include species we would interpret as only semi-infaunal, byssally attached nestlers. Some Gastrochaenacea have reduced the length of the nymph from that of their permophorid ancestors. *Spengleria rostrata* (Spengler), from St Thomas, has a broad short nymph, whereas *Gastrochaena mytiloides* (Lamarck) from Mauritius has an elongate ligament set in a notch along the top of the hinge, suggesting that the length of the ligament may be highly adapted to slight differences in life style and might not always reflect phylogeny.

The Edmondiacea also have a relatively long ligament nymph (Runnegar & Newell 1974) which may be a primitive

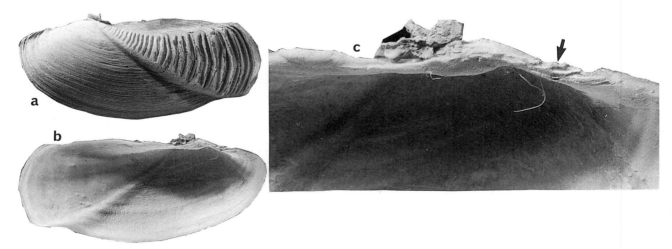

Fig. 2 *Spengleria rostrata* (Spengler). Recent, St Thomas I., Caribbean. BM (ZD) unnumbered (Cuming Collection, *ex* Dr Hornbeck). Figs 2a, 2b, exterior and interior view of left valve, slightly enlarged. Fig. 2c, interior view of dorsal margin of left valve, showing the development of ridges anterior to the umbones due to the emplacement of anterior fused periostracum (arrowed), ×6.

character for that superfamily. Some Sanguinolitidae, as well as the majority of species of *Pholadomya*, develop a marked permanent posterior gape for the siphons. By comparison with living taxa, such large gapes are invariably accompanied by fused siphons of type 'C' (Yonge, 1957). In *Chaenomya* the gape extends along the posterior part of the dorsal margin. The escutcheon is partly lost and it is clear that the dorsal margins behind the nymphs were not continuously joined by periostracal ligament. We take this to be an advanced character, associated with largely or at least partly retractile siphons.

In *Spengleria rostrata* the valves are joined by periostracum anterior to the umbones. In dead shells the anterior dorsal margin has a chalky texture with layers of periostracum along the shell margin (Fig. 2). This clearly resembles, and we consider it to be analogous to, the structure figured by one of us (Astafieva-Urbaitis 1964) in a species of *Allorisma*, where it was suggested that they may be anterior teeth.

SYSTEMATIC DESCRIPTIONS

Abbreviations. Specimens in the following institutions are referred to in the text and figure captions with the following abbreviations:

BGS   – British Geological Survey, Keyworth, Nottingham.
BM   – The Natural History Museum, Cromwell Road, London (formerly the British Museum (Natural History)).
BMR   – Bureau of Mineral Resources and Mines, Canberra.
EMP   – École des Mines, Paris (now at Université de Lyon, France).
GSI   – Geological Survey of India, Calcutta.
NMI   – National Museum of Ireland, Dublin.
SM   – Sedgwick Museum, Cambridge.
USNM   – United States National Museum, Washington.
MNHN   – Musée Nationale d'Histoire Naturelle, Brussels.

Subclass ANOMALODESMATA Dall, 1889

The Upper Palaeozoic taxa included in this subclass are listed on p. 92. We describe here representatives of the Edmondiacea and Pholadomyacea. What we interpret as primitive characters within the subclass include an equivalve nacreoprismatic shell composed entirely of organic material and aragonite, joined across the dorsum by a three-layered, opisthodetic, parivincular ligament mounted on narrow but distinctive nymphs. More or less isomyarian adductor muscles are joined by an entire pallial line. We are uncertain whether the most primitive Anomalodesmata possess simple hinge teeth or have none. We regard the possession of spicules of aragonite on the shell surface set within the periostracum as a synapomorphy of the Anomalodesmata, but we are prepared to modify this view if a sister group of the extant representatives is recognized which did not develop this feature. We consider it unlikely that the view expressed by Carter & Aller (1975), that these periostracal spicules are the homologue of similar structures in chitons and therefore a primitive character of the Mollusca, is correct. Most Anomalodesmata have somewhat elongate shells, which we consider to be primitively infaunal, often with modification of the form of the posterior margin, which suggests that the inhalant water current of the mantle cavity was posteriorly placed. This may not be the case in primitive edmondiaceans. A subumbonal sulcus is present in many anomalodesmatans and may indicate the anterior limit of ventral mantle fusion.

Pojeta (1971) has excluded the family Orthonotidae from the Anomalodesmata, suggesting first that they should be allotted to a separate Order Orthonotoida (Pojeta 1978), but he later suggested an affinity with the Mytiloida (Pojeta, Zhang & Yang 1986). On all occasions, and with support from Runnegar (1974), the similarity between the Orthonotidae and the living Solenacea was stressed by these authors. A separate (unpublished) study (N.J.M. *in litt.*) supports an alternative view that the Solenacea are more closely related to the Tellinacea, and both superfamilies probably arose from an ancestor currently classified with the Tancrediidae. We suggest that the *Solen*-like shape has arisen at least four times throughout bivalve evolution; once within primitive Ordovician forms of uncertain affinity, once in the Orthonotidae, once in the Quenstedtiidae and at least once in the Solenacea. We consider all these cases to be due to convergence. The significance of the Orthonotidae to the classification relates to our inablility to decide, on the presently available evidence, whether the Devonian to Triassic family Solenomorphidae is more closely related to early Devonian Sanguinolitidae such as '*Leptodomus*' *acutirostris* (Sandberger) (Beushausen, 1895: pl. 24, figs 8–10) or to *Orthonota*. Bittner (1895: pl. 1) illustrated a series of species of *Solenomorpha* from the Carnian of northern Italy which grade insensibly into what seem to be the earliest representatives of the Cuspidariidae. Interpretation of their ancestry will indicate whether or not the Cuspidariidae are properly placed in the Anomalodesmata. The shell shape, differentiation of a corselet and subumbonal sulcus, all typical of species of *Orthonota*, are not characters of the Mytilacea. However, until a well-preserved hinge is described for that genus, its systematic position remains debatable.

We include provisionally a new Subfamily Promacrinae (p. 93) within the Solenomorphidae. *Promacrus* is superficially similar to the living arcacean *Litharca*, but apparently does not have an arcid hinge. Many of the species have an opisthodetic parivincular ligament with the umbones well to the posterior, and are similar in this respect to the Solemyidae. However, a specimen of *Promacrus* in the United States National Museum labelled *Promacrus undatus* Ulrich MS, from the Lower Cuyahoga Shale of northern Ohio, has a clearly preserved anterior adductor scar, which does not impinge upon the body attachment scars in the way peculiar to the Solemyidae. In addition we take the condition of *Promacrus websterensis* as originally illustrated by Weller (1899; 34, pl. 2, figs 2–7; pl. 3, fig. 1), where the umbones are more medially placed, as primitive for the genus and subfamily, and therefore less similar to the solemyid shape. We provisionally place the Promacrinae in the Solenomorphidae but feel this view may have to be modified when the hinge of *Promacrus* is adequately described.

The Prothyridae are another difficult family to place with certainty, but in Driscoll's illustration of the surface sculpture (Driscoll, 1965: pl. 11, figs 1–9), the fine radial striae resemble similar structures in a number of Mesozoic Anomalodesmata. Although carbonate spicules are not preserved in *Prothyris* a relationship to the Anomalodesmata is indicated.

Superfamily EDMONDIACEA King, 1850

The Edmondiidae include three oval genera, *Edmondia*, *Scaldia* and *Cardiomorpha*, which are more or less isomyarian with an entire pallial line, and a more elongate genus *Allorisma*. *Edmondia*, *Scaldia* and *Allorisma* share an internal rib below the hinge plate which is not usually well developed in species of *Cardiomorpha*. The Edmondiacea share only primitive characters with the other Anomalodesmata, except for the rare occurrence of periostracal spicules. All the other characters we are able to recognize, we would expect to be primitive in early members of the Heteroconchia. The superfamily is placed within the Anomalodesmata largely by tradition. However, by assuming the periostracal spicules are a synapomorphy for the Anomalodesmata as a whole we support their inclusion in this subclass. At present we find no irrefutable evidence for the occurrence of edmondiaceans before the latest Silurian, but we suspect that earlier representatives will be either recognized or confirmed.

The rounded edmondiid shell with its regularly curved pallial line is consistent with a mantle cavity lacking a specialized channelling for the inhalant current. We consider the more elongate form and more intricate musculature of *Allorisma* to be advanced features. This lack of channelling of the inhalant current might have been similar to that of primitive living Veneroida such as *Astarte* and nuculoids such as *Nucula*, and may well have been primitive for the Lucinoida. Unlike representatives of the latter order, no edmondiid or conceivable relative discussed below, megadesmid, mactromyid or poromyid, has the hypertrophied anterior adductor scar of the Lucinacea, which is clearly visible in the earliest certain lucinacean, *Ilionia* (from the Silurian of northern Europe).

We include two Palaeozoic families in the Edmondiacea, the Edmondiidae and the Megadesmidae. The Edmondiidae are found mainly in strata of Devonian to Permian age, laid down in temperate and warm seas, now in the northern hemisphere, whereas the Megadesmidae mainly occur in the cold and cold-temperate waters of the southern hemisphere of Permian times. De Koninck (1877–8) distributed his new species from the Permian of Australia, now recognized as belonging to *Megadesmus*, among his own genera *Edmondia* and *Cardiomorpha*. Although the megadesmid genera from Australia have a characteristic shape, a relatively larger size, immensely thicker shells and more robust ligaments than Carboniferous species of *Edmondia* and *Cardiomorpha*, the similarity implied by de Koninck is real. The relationship has been widely discussed and has been summarized by Runnegar (1967: 29). We have compared the shell shape and hinge structure of a young specimen of *Megadesmus grandis* (Dana) and the musculature of several specimens of *Astartila intrepida* (Dana), both from the Illawara District of New South Wales, Australia, with the same features of a well-preserved Lower Namurian English specimen of *Cardiomorpha obliqua* Hind, and find no significant differences. Unlike *Edmondia, Scaldia* and *Allorisma*, *Astartila* and *Megadesmus* have no significant internal rib on the hinge plate. We know of no taxon more similar to these megadesmids than *Cardiomorpha obliqua*, and interpret that species as the closest known relative to the Megadesmidae. We have noted the superficial similarity in shape and sculpture between *Cardiomorpha* and the Jurassic genus *Ceratomya*. Both have a similar size, globose form and comarginal ribs, which has led Runnegar to speculate (1974: text-fig. 3) that *Ceratomya* evolved from the edmondiids and occupied a deeper burrowing habitat, signified by the acquisition of a pallial sinus. *Ceratomya* also developed an internal ligament by overlap of the left nymph by the right valve, and further has homogeneous rather than nacreoprismatic shell structure. Our greatest misgiving, however, is in the shell morphology of the oldest genus we attribute to the Ceratomyidae and which we would interpret as the primitive

morphology of that family. This is the Rhaetian to Hettangian genus *Pteromya*, which is more elongate and less gibbous than any *Cardiomorpha* or *Edmondia*. At present we are impressed by the external similarity between the sculpture of *Pteromya* and that of Middle Jurassic species of Cuspidariidae, and we conclude that the similarity between the gibbous shells of *Ceratomya* and *Cardiomorpha* is a consequence of convergence.

The question still arises as to whether the Edmondiacea survived beyond the end of the Palaeozoic. At an early stage we thought that the Corbulacea (Mesozoic to Recent) were derived from the Megadesmidae (outlined in Taylor *et al.* 1973, final chapter). However, we now favour a relationship between the Corbulacea and the Crassatellacea, particularly the Triassic family Myophoricardiidae. We consider the similarity of hinge structure between the Corbulidae and the Myidae to be best interpreted as a case of convergence. It is hoped to enlarge on this hypothesis in a later paper.

It has been suggested by Runnegar & Newell (1974) that the late Triassic genus *Ochotomya* Kiparisova *et al.* 1966 may have evolved from a megadesmid ancestor. It is possible that *Ochotomya* is an early representative of the Poromyacea. We can find no authentic poromyacean earlier than the Late Campanian; we include the *Liopistha* group in the Cardiacea on the basis of their cardinal teeth, shell structure and muscle scars. In spite of the considerable time gap between the Late Triassic and the Campanian we have so far been able to suggest no more plausible relationship for the Poromyacea. However, *Bowlandia* sp. (Fig. 45) is an equally plausible relative for *Ochotomya*. The basis for separation of the Edmondiacea is discussed under Pholadomyacea. Here we note the considerable convergence between the genus *Allorisma* of the Edmondiidae and genera of the sanguinolitid Subfamily Undulomyinae. They are distinguished by the pattern of the accessory musculature and the presence of the hinge plate rib which appears to be a synapomorphic character of a number of Edmondiidae, although it may be lost during the course of later evolution. The possibility arises that the Mesozoic genus *Mactromya* and its relatives are descendants of the genus *Edmondia*, and that the Edmondiacea as a whole are better placed in the Heteroconchia. We are uncertain whether the subumbonal hinge teeth present in the edmondiid genus *Scaldia* and the Mesozoic genera *Sphaera* and *Schafhaeutlia* are an advanced or primitive character for the group. We have been unable to ascertain the shell structure of Palaeozoic Edmondiacea. In the Upper Pennsylvanian of Texas an unnamed species occurs in a preservation similar to that which has yielded shell structure information in some other taxa; only the growth laminae were clearly preserved. It is possible that the inner layer was nacreous but it is not sufficiently well preserved for us to be certain. Runnegar (1967: pl. 6, fig. 12) has illustrated what appears to be the nacreo-prismatic shell of the genus *Megadesmus*. The Lower Jurassic *Mactromya cardioides* (Phillips), from the Lower Pliensbachian, Luridum Subzone of Blockly, Gloucestershire, England, shows clear aragonite crossed lamellae and some finer structure that was apparently of amorphous type. This shell structure does not conflict with our view that the Mactromyidae may be closely related to the Lucinacea. Further work is necessary to decide whether the similarity of the Edmondiidae to the Mactromyidae is the result of convergence or descent.

Family **EDMONDIIDAE** King, 1850

Genera referred here to this family are *Edmondia, Allorisma, Scaldia* and *Cardiomorpha*, which all possess a hinge plate reinforced internally by a ridge or lamellar plate projecting into the shell cavity; it is equivalent to the 'internal cartilage plate or ossicle' (Wilson 1960, Waterhouse 1969a, Runnegar & Newell 1974). Earlier, one of us (Dickins 1963) had suggested that the Megadesmidae might be included in the family Edmondiidae, but this relationship has been discussed by Runnegar & Newell (1974), who showed that *Allorisma, Scaldia* and *Cardiomorpha* also had an internal lamellar plate as well as having other characters in common with *Edmondia*. On the basis of this information and further data presented here we also conclude that the Edmondiidae and the Megadesmidae should be recognized as separate families. Edmondiids from western Europe such as *Cardiomorpha obliqua* Hind (1898: 263) and *Edmondia lyellii* Hind (1899: 300) are the most similar to typical Australasian genera *Megadesmus, Pyramus* and *Astartila*. We imagine the ancestry of the Megadesmidae to have been from taxa such as these. *Cardiomorpha obliqua* in particular is thick-shelled for an edmondiid. Although its hinge margin is quite thick, the internal rib is only just distinct and its adductor and accessory muscle scars are rather similar to those of the megadesmids. We suspect that the species of *Vacunella, Australomya* and *Myonia* from eastern Australia are true anomalodesmatids, which are not necessarily closely related to the megadesmids. Periostracal spicules are not usually preserved on the shell surface of Edmondiacea, but Runnegar (1965: pl. 13, fig. 9) clearly illustrates their presence in *Megadesmus gryphoides* (de Koninck). We are unable to say whether they originally occurred on other taxa within this superfamily.

We follow Astafieva-Urbaitis (1964) and Runnegar & Newell (1974) in separating *Allorisma* from *Edmondia* and add some further information on the musculature of *Allorisma*. *Scaldia* can be separated, as it possesses a hinge tooth, and *Cardiomorpha* includes slender to inflated circular shells with inrolled umbones. The Edmondiidae were apparently shallow burrowers and did not develop in the same area as the Megadesmidae. The Edmondiidae are found mainly in strata laid down in temperate and warm seas of the northern hemisphere, whereas the Megadesmidae mainly occur in the cold and cold temperate waters of the southern hemisphere.

Genus *ALLORISMA* King, 1844: 315
Fig. 3

TYPE SPECIES. *Hiatella sulcata* Fleming (1828: 462) by subsequent designation of King (1850: 196, footnote 6) (not *Cardiomorpha sulcata* de Koninck, 1842, which is not an *Allorisma*).

Wilson (1960: 114) and Newell (1969: N818) considered

Fig. 3 *Allorisma sulcata* (Fleming). Carboniferous. Figs 3a–b, Lower Namurian, Main Limestone, Stanhope, Northumberland; BM PL5000, Trechmann Collection. Fig. 3a, anterior of right valve with anterior adductor AA, and two anterior pedal-body attachment scars APM1 and APM2; Fig. 3b, posterior dorsal area of left valve with attachment scar or posterior pedal retractor PPR. Figs 3c–d, Viséan, Ballasalla, Isle of Man; BM L45456, dorsal and side views, ×1.

Sanguinolaria sulcata Phillips 1836 to be the type species of *Allorisma* by original designation of King (1844: 313). In his original description of *Allorisma*, however, King mentions several species and only in the letter of introduction to his paper does he refer to the new genus '*Allorisma* for species represented by *Sanguinolaria sulcata*. Ph'. This does not seem to represent an explicit designation of type species, whereas in 1850 King made a definite designation. Wilson (1960: 112) chose a lectotype for *Hiatella sulcata* Fleming from amongst Fleming's specimens, and presented evidence that *Sanguinolaria sulcata* Phillips 1836 should be regarded as a synonym of *Hiatella sulcata* Fleming 1828.

SYNONYMS. *Edmondiella* Chernychev, 1950: 74 (type species, *Sanguinolaria sulcata* Phillips 1836 by original designation). *Edmondia* King 1850 *pars* (1850: pl. 20, figs 1–2) (not *Edmondia* de Koninck, 1844).

DESCRIPTION. Thin-shelled, elongate oval, with umbones situated distinctly towards the front. Rather evenly rounded over the surface of the shell and lacking a distinct escutcheon or lunule. Ornament of well developed rounded rugae more or less parallel to the external margin. In internal impressions a smooth area is marked off by the anterior adductor. Has a distinct internal ridge or lamellar plate as in other Edmondiidae. Hinge lacks teeth and apparently with an external opisthodetic ligament. Anterior adductor muscle scar moderately well marked at right angles to margin in front of umbones rather than vertical. Two other separate muscle attachment marks are associated with the adductor (Fig. 3). The longer scar runs parallel to the front part of the dorsal margin and behind has a distinct buttress (clavicle). At the dorsal end of the buttress is the rounded mark of another muscle. The two smaller muscles perhaps represent the pedal protractor and retractor. The posterior adductor muscle is poorly marked but a posterior pedal retractor is visible towards the back of the dorsal margin. Delicate lines radiate from the umbones of external impressions.

REMARKS. The description is largely based on the type species. The elongate shape seems sufficient to distinguish *Allorisma* from *Edmondia*, which, in addition, commonly has lamellate shell ornament not so far recorded in *Allorisma*. Although the anterior accessory muscles are similar in *Edmondia* (Runnegar & Newell 1974) and *Allorisma*, a distinct buttress is not recorded in *Edmondia* nor the muscle impression at the dorsal end of the buttress.

[*Note: Allorisma regularis* de Verneuil (1845: 298; pl. 19, figs 6a–b; pl. 21, figs 11a–b) was attributed by de Verneuil to King out of courtesy. It was, however, never described by King but is referred to by him (King 1850: 196) as described by de Verneuil. The two specimens figured by de Verneuil represent two species. We here designate the specimen figured in de Verneuil, 1845: pl. 19, figs 6a–b as the lectotype, to safeguard the usage of Astafieva-Urbaitis (1962), who was apparently the first revising author. This specimen belongs to the genus *Wilkingia*. The other specimen (de Verneuil 1845: pl. 21, figs 12a–b) has a lamellar plate and belongs to *Allorisma* as used in this paper, and as appreciated by King in his footnote. It is here considered to be a synonym of *Allorisma sulcata*.]

Superfamily PHOLADOMYACEA King, 1844

Although the family name Pholadomyidae was used by Gray (1847: 194) (Newell 1969: N818, Runnegar 1974: 425), it was used earlier by King (1844: 315). The name Pholadomyacea is assigned by Newell (1965: 21) to Fleming (1828), but this is apparently an error because Fleming (1828: 424) referred *Pholadomya* to the Cardiadae (*sic*).

In the *Treatise*, Newell (1969) separated the Edmondiidae as a single family in the superfamily Edmondiacea, but Runnegar & Newell (1974) place the family in the Pholadomyacea. In the latter case the name Edmondiacea would become redundant, as also would Grammysiacea, the Grammysiidae being placed in the Pholadomyacea, which has priority over both names. The superfamily has thus become very broad and we have misgivings about including some Palaeozoic genera, which, although generally related to *Pholadomya*, are rather distinctive. Runnegar & Newell and our present work emphasize the significance of the lamellar plate in the Edmondiidae. We consider the muscle attachment associated with the plate would give greater stability and strength to the valves in burrowing. Apparently this feature is a long-standing one, as the lamellar plate is also developed in a Welsh Lower Devonian species, ?*Edmondia* sp. from the Tilestones at Capel Horeb (BM LL31477). We propose to emphasize the significance of this feature by recognizing the Superfamily Edmondiacea to include the Family Edmondiidae.

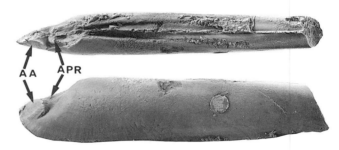

Fig. 4 *Solenomorpha minor* (M'Coy). Lower Carboniferous, Viséan, Yeat House Quarry, Cumberland. BM L47699, Hind Collection (figd Hind, 1904: 159; pl. 22, fig. 3); views of steinkern from the top and left side, showing anterior adductor AA, and anterior pedal retractor APR. ×1.

We place the Sanguinolitidae and the Permophoridae in the Superfamily Pholadomyacea, although this is not altogether satisfactory because these two families are distinct from the rest of the superfamily. Alternatively, to separate a Superfamily Grammysiacea for these two families would seem even less satisfactory. The later representatives of *Grammysia* itself appear to be a group of round-shelled, non-siphonate species for which we can find no evidence of survival beyond the end of the Devonian. We would be reluctant to propose another superfamily without further work and review of pre-Carboniferous and post-Permian bivalve faunas, because the lineages we are discussing and using as a basis of family taxonomy occur in the Carboniferous and later faunas, and this information will allow the establishment of more satisfactory systematics.

Family **SANGUINOLITIDAE** Miller, 1877

SYNONYM. Caelonotidae M'Coy, 1855, an invalid name according to ICZN Art. 11e (not Coelonotidae).

REMARKS. On the basis of a better understanding of *Sanguinolites* and the assignment of species hitherto placed in *Sanguinolites* to *Myofossa*, *Pleurophorella* and *Gilbertsonia*, a more satisfactory definition of *Sanguinolites* is possible. The family contains more or less transversely elongate shells with a lunule and a distinct, flattish escutcheon. The ligament is lodged in an opisthodetic groove at the front point of the escutcheon, with a small slender nymph. Earlier, Dickins (1963) considered that the flattish elongated area (escutcheon) behind the umbones was a ligament area, but fresh work has shown that the bulk of the ligament, the spring part, is lodged in a groove at the front end of the area (Fig. 11d). It is possible, however, that the periostracal ligament was continuous with that covering the shell surface in this region and that it continued posteriorly to the distal end of the escutcheon (Yonge's (1957) fusion layer, see p. 54, discussion on hinges). Shells with granular surface spicules are found in this family. We divide the family into Subfamilies Sanguinolitinae, Undulomyinae, Pholadellinae, Alulinae, Chaenomyinae and Vacunellinae. The similarity of escutcheon and posterior area or corselet exhibited between the Sanguinolitidae and the permophorid *Pleurophorella*, which led Hind (1900) to include both in his interpretation of the genus *Sanguinolites*, has suggested to us a common origin for the two.

Subfamily **SANGUINOLITINAE** Miller, 1877
[*nom. trans.* herein, *ex* Sanguinolitidae Miller]

In this subfamily are placed the genera *Sanguinolites*, *Myofossa* and *Gilbertsonia* along with the several genera listed on p. 94. These forms were apparently adapted to a shallow infaunal environment. *Sanguinolites* especially retained unspecialized features. *Myofossa*, on the other hand, has characteristic features of the escutcheon which at some future time may serve as the basis for recognizing a separate subfamily. The family Permophoridae, to which *Sanguinolites* seems closely related, has specialized by the development of subumbonal hinge teeth for a shallow infaunal environment, and from time to time species became specialized for an endobyssate habitat.

Genus *SANGUINOLITES* M'Coy, 1844: 47

TYPE SPECIES. *S. discors* M'Coy (1844: 49; pl. 8, fig. 4), subsequently designated by Stolizcka (1871: xix, 270) (= *S. angustatus* (Phillips) (1836: 208; pl. 5, fig. 2)). Hind (1900: 367) concluded that M'Coy's type of *S. discors* was a young specimen of *S. angustatus* (Phillips, 1836), and our examination of the type specimens and other material confirms that *S. discors* is a junior subjective synonym of *S. angustatus* (Phillips) 1836.

DESCRIPTION. Elongate with umbones well anterior of midpoint. Well differentiated posterior dorsal area, with three radiating ribs present. The top one delimits an elongate marked escutcheon, there is a medially placed one, and a third, running to the posterior ventral margin, delimits the area itself and also the subvertical siphonal margin. The area

is ornamented with comarginal growth laminae, some of which may form thin ribs. Ventral margin curved, flank ornamented with distinct rounded ribs separated by wider interspaces. Dorsal margin long and straight behind umbones set in a distinct escutcheon. A ligament groove extends behind the umbones, separating a slender but distinct nymph. The straight dorsal margins behind the nymph were probably joined by periostracal ligament (Fig. 6). Moderately impressed anterior adductor scar, rounded, with a deeply inset anterior pedal retractor at its dorsal umbonal edge extended more or less parallel to the anterior umbonal ridge, at about 45° to the cardinal margin and not at right angles as in the Permophoridae. No buttress visible.

Posterior adductor scar barely discerned. Pustules visible above ribs towards the anterior ventral margin in *S. costatus* Meek & Worthen 1869 (Upper Pennsylvanian, Texas). There are few records of pre-Carboniferous species of *Sanguinolites*, although we consider '*Leptodomus' acutirostris* Sandberger as illustrated by Beushausen (1895: pl. 24, figs 8–10) from the L. Devonian of the Rheinland to belong here or to a very closely related genus.

Sanguinolites angustatus (Phillips, 1836)   Figs 5a–d

1836   *Sanguinolaria angustata* Phillips: 208; pl. 5, fig. 2.
1844   *Sanguinolites angustatus* (Phillips); M'Coy: 48.
1844   *Sanguinolites discors* M'Coy: 49; pl. 8, fig. 4.
1900   *Sanguinolites angustatus* (Phillips); Hind: 366–8; pl. 40, figs 1–6.
?1910   *Sanguinolites simulans* Girty: 224.
?1969   *Sanguinolites simulans* Girty; Pojeta: 16; pl. 3, figs 3–5.

TYPES. BM PL4272, the lectotype of *Sanguinolaria angustata* Phillips, from 'Bolland', Yorkshire/Lancashire border, Lower Carboniferous, Viséan. National Museum of Ireland, the lectotype (selected herein) of *Sanguinolites discors* M'Coy (1844: pl. 5, fig. 2). USNM 155895 is the holotype of *Sanguinolites simulans* Girty; Pojeta (1969: 16) stated that this is the only original specimen of Girty.

OTHER MATERIAL. BM 97184 and BM L47471, 2 paralectotypes of *S. angustata* from Poolvash, Isle of Man. BM 52029 from Clifton, near Bristol. One paralectotype (Nat. Mus. Ireland, not numbered) of *Sanguinolites discors*, from the Arenaceous Shales at Bruckless, Dunkinelly, N.W. Ireland.

REMARKS. We have examined and refigured (Fig. 5c) Phillips' type specimen of *Sanguinolaria? angustata* and M'Coy's type specimen of *Sanguinolites discors*. Our examination confirms the conclusion of Hind (1900: 366) that *S. discors* is a synonym of *S. angustatus*.

Sanguinolites costatus (Meek & Worthen, 1869)   Fig. 6

1869   *Allorisma costata* Meek & Worthen: 171.
1873   *Allorisma costata* Meek & Worthen; Meek: 585–6; pl. 26, fig. 15.

MATERIAL EXAMINED. USNM 1506, two specimens from the Upper Carboniferous, Pennsylvanian Cisco Formation, near Jacksboro, Texas.

REMARKS. This Upper Carboniferous species is very similar to *S. angustatus* in both shape and sculpture. It may be distinguished by its wider-spaced comarginal ribs. Small numbers

Fig. 5 *Sanguinolites angustatus* (Phillips). Figs 5a–b, Arenaceous shale, Bruckless, Dunkineely, County Donegal, Ireland: NMI, Griffiths Collection; Fig. 5a, **lectotype** (selected herein), and Fig. 5b, paralectotype (on same piece of rock), of *S. discors* M'Coy, both ×3. Fig. 5c, Lower Carboniferous, Viséan, Bolland, Yorkshire; BM PL4272, Gilbertson Collection, lectotype of *Sanguinolites angustatus*, side view, ×1·5. Fig. 5d, Lower Carboniferous, Clifton, Bristol; BM 52029, ×1·3.

Fig. 6 *Sanguinolites costatus* Meek & Worthen. Locality *24827, Upper Carboniferous, Pennsylvanian, Cisco Formation, near Jacksboro, Texas; USNM Dr* 1506, Renfro Collection; Fig. 6a, top view; Fig. 6b, oblique dorsal view; Fig. 6c, slightly oblique view of left side showing the ligament nymph of the right valve; all approx. ×2.

of periostracal spicules were observed on the lower anterior part of the flank below the comarginal ribs. Our interpretation of this is that their function of maintaining the position of the shell in the sediment had been taken over by the comarginal ribs. The value of this species relates to its ligament attachment area being unequivocally well preserved. It has relatively long, low, slender nymphs separated from the outer escutcheon surface by a deep ligament groove.

?*Sanguinolites argutus* (Phillips, 1836)   Figs 7a–b

1836   *Cucullaea arguta* Phillips: 210; pl. 5, fig. 20.
1897   *Cucullaea arguta* Phillips; Hind: 174.
1900   *Sanguinolites argutus* (Phillips) Hind: 368–9; pl. 40, figs 15–16.

HOLOTYPE. BM 97155, Lower Carboniferous, Viséan, 'Bolland', Yorkshire/Lancashire Border, England; Gilbertson Collection.

DISCUSSION. ?*Sanguinolites argutus* is a much more tumid shell than *S. angustatus*. The posterior area or corselet is marked by a very sharp angularity, almost a carina; this leaves the area diverging from the plane of commissure at a fairly high angle until it approaches the siphonal margins. Hind described this species as one of the rarest in the Carboniferous. It may, however, prove to be an important species because it resembles *Myonia carinata* from the Permian of eastern Australia in its morphology. As yet we are uncertain whether this is due to convergence or whether it indicates a natural relationship.

Fig. 7 *Sanguinolites argutus* (Phillips). Lower Carboniferous, Viséan, Bolland, Yorkshire, England; BM 97155, Gilbertson Collection, holotype; Fig. 7a, right side, Fig. 7b, dorsal view; both ×3.

Genus *MYOFOSSA* Waterhouse, 1969*b*

TYPE SPECIES. *Myonia subarbitrata* Dickins, 1963 (p. 48; pl. 5, figs 2–12) by original designation.

REMARKS. In attempting to decide on the correct generic name for this taxon we have considered the following nominal genera in addition to *Myofossa*:

Sedgwickia M'Coy, 1844, type species *S. attenuata* M'Coy, 1844, by subsequent designation of Stoliczka, 1871 (Fig. 8).

Palaeocorbula Cowper Reed, 1932, type species *P. difficilis* Cowper Reed, 1932, by monotypy (Fig. 15).
Grammysiopsis Chernychev, 1950, type species *Grammysiopsis irregularis* Chernychev, 1950, by original designation.
Ragozinia Muromzeva 1984; type species, *Ragozinia dembskajae* Muromzeva & Guskov *in* Muromzeva, 1984, by original designation.

DISCUSSION. The type specimen of *Sedgwickia attenuata* M'Coy (1844: 62; pl. 11, fig. 39; refigured by Hind, 1899: pl. 27, fig. 8) has been kindly lent by Dr Colm E. O'Riordan, formerly of the Natural History Division of the National Museum of Ireland, Dublin. It is refigured here (Fig. 8). *S. attenuata* is the type species of *Sedgwickia* by subsequent designation of Stoliczka (1871: xix, 271); Chernychev's (1950: 33) designation of *Sedgwickia gigantea* M'Coy is invalid. The generic name was used by Hind (1899) and Runnegar (1974: 932), and the species to which Runnegar refers are included here in *Myofossa*. However, Hind included a variety of species, including some that we would ascribe to the trigoniacean family Schizodidae. We are unable to interpret the type species, *S. attenuata*, from its holotype (Fig. 8), which is small and badly crushed. The ribbing is poorly preserved and there are no details of the hinge or musculature. In particular the characteristic form of the escutcheon which would allow us to refer it in the present genus is absent or not preserved. Recollecting from the type locality might show that M'Coy's species is a synonym of *Sanguinolites variabilis* M'Coy of Hind (not *Cosmomya variabilis* (M'Coy)). This species is described here as *Myofossa hindi* sp. nov. With the present state of knowledge the binomen *Sedgwickia attenuata* M'Coy, 1844 should be rejected as a *nomen dubium*.

Fig. 8 '*Sedgwickia attenuata*' M'Coy (*nomen dubium*). Lower Viséan, Arenaceous Shale [of Yellow Sandstone Group], River Bannagh, Drumcurren (near Kesh, County Fermanagh), Northern Ireland; NMI, Griffiths Collection, holotype; view of crushed composite mould of left valve, ×3.

Chernychev (1950) included two species in his new genus *Grammysiopsis*: the type species, *G. difficilis*, and *G. kazachstanensis* n. sp. His own '*Grammysioidea*' *welleroides* (Cherychev, 1950: pl. 7, fig. 68 only) may be a synonym of the latter. His type species is rather poorly illustrated and we were unable to say with certainty whether this is a *Myofossa*. The peculiar grouping and turning of the comarginal ribs towards the posterior area suggest to us that this should be considered as a separate genus, at present not described outside the USSR. However, the second species, *G. kazach-stanensis* (Cherychev, 1950: pl. 6, figs 53a–d, 54a–b) is almost identical to *Myofossa hindi* described here. Other species doubtfully included by Chernychev, *G. donaica* and *G. obscura*, may be synonyms of *Cosmomya variabilis* (M'Coy). The problem appears to have been solved by Muromzeva (1974), who illustrated a number of species of *Grammysiopsis*, ranging from the Viséan to the Carboniferous–Permian boundary, that have the same characteristic twist to the ribs as Chernychev's type species. From this material from Kazakhstan and the Soviet Arctic we can see that *Grammysiopsis* may be easily distinguished from *Myofossa* by the enormous size of its posterior gape. *Grammysiopsis* appears to be a distinct offshoot of *Myofossa* which has developed much more substantial siphons, presumably of type 'C', that has so far not been recognized outside the Soviet Union.

Cowper Reed's (1932) genus *Palaeocorbula* is based on a single specimen (Figs 14a–c). It is smaller than most species of *Myofossa* and shows no sign of the posterior attenuation. This marked difference in shape leads us to accept it as a distinct genus. It is possible, however, that the specimen is deformed and the similar pattern of ribbing on the anterior and central part of the flank may necessitate the future synonymizing of the two taxa when more material becomes available.

SUBDIVISIONS OF *Myofossa*. At present we recognize two subgenera of *Myofossa*: *Myofossa* s. str. and *Ragozinia*. Species of *Myofossa* from the British Lower Carboniferous have been described as *Sanguinolites* by Hind (1900). All species of *Myofossa* are more convex, relatively shorter and do not have the distinctly delimited and sculptured corselet possessed by *Sanguinolites*. Driscoll (1965) included the species *Myofossa omaliana* (de Koninck), which occurs in both NW Europe and the U.S.A., in *Grammysia*. However, the latter genus has a less attenuate posterior shell, a less well defined siphonal area, usually no gape and a completely simple pallial line.

DIAGNOSIS. Oval in shape. Distinguished mainly by the features of the escutcheon. This consists of an internal heart-shaped area in which the ligament is lodged in a groove immediately below the umbo; in turn this is bound externally by a distinctly marked-off escutcheon which is again bounded externally by the umbonal ridge. Shell generally gaping at the rear with a more or less well developed sulcus running from the umbo towards the mid-part of the ventral margin. Flank usually bearing distinct, low, comarginal ribs, often fewer in number in a distinct anterior area, and more on the main part of the flank.

REMARKS. A number of Carboniferous species share the characteristic features of *Myofossa*. The Carboniferous species have a small ligament nymph, but this hardly seems sufficient to place them in a different genus. They include *Sanguinolites omalianus* de Koninck, 1842, *S. costellatus* M'Coy, 1851a, and *Myofossa hindi*, which can be distinguished from *Sanguinolites* proper, as well as some other genera, by their more oval shape. From other Megadesmidae and Edmondiidae they are distinguished by the distinctive features of their escutcheon.

During the course of this work we first thought that *Myofossa* might be placed with the Megadesmidae. However, the nature of the escutcheon and the persistence of this feature in bivalves from the middle of the Palaeozoic to the Mesozoic has caused us to regard this feature as of considerable importance. We have, therefore, assigned *Myofossa* to the Sanguinolitidae which have similar characteristics of the escutcheon.

Species of *Myofossa* have a variety of shell form comparable with living species of *Laternula* Röding, 1798. At present it does not seem to us that the earlier species of the Laternulidae and the Thraciacea as a whole are derived from them; we consider it more likely that the similarity between Upper Palaeozoic *Myofossa* and Recent Laternulidae is a matter of partial convergence.

Subgenus *MYOFOSSA (MYOFOSSA)*

Myofossa (Myofossa) subarbitrata (Dickins, 1963)
Figs 9a–f

1963   *Myonia subarbitrata* Dickins: 48–9; pl. 5, figs 2–12, 22.
1969b   *Myofossa subarbitrata* (Dickins) Waterhouse: 32, 66.

MATERIAL. The type material from the Lower Permian, Nura Nura Member of the Poole Sandstone Formation, of the Canning Basin, Western Australia, all in the Bureau of Mineral Resources and Mines, Canberra, Australia.

DISCUSSION. New illustrations are given here (Figs 9a–f) to show the striking similarity between the present species and British Carboniferous species of *Myofossa*.

Myofossa (Myofossa) hindi sp. nov.   Figs 10a–e

1851a   *Sanguinolites variabilis* M'Coy: 174 (pars).
1855   *Sanguinolites variabilis* M'Coy; M'Coy: 508; pl. 3f, fig. 7 only.
1900   *Sanguinolites variabilis* M'Coy; Hind: 379; pl. 44, figs 3–8 only.
?1900   *Sanguinolites variabilis* M'Coy; Hind: 379; pl. 44, fig. 1 only.
?1900   *Sanguinolites v-scriptus* Hind: 382; pl. 42, figs 5, 5a only.

HOLOTYPE. BM L47511, from the Viséan Redesdale Ironstone (Figs 10a–d) (also figured by Hind, 1900: pl. 44, fig. 3.)

PARATYPES. BM L47512–47516, the specimens figured by Hind (1900: pl. 44, figs 4, 4a, 5, 6, 7 & 8) and BM L3231 (Fig. 10e here), all from the Redesdale Ironstone; Sedgwick Museum, Cambridge, the type of M'Coy's ovate variety of *Sanguinolites variabilis*, from the Carboniferous Limestone of Lowick, Northumberland.

DIAGNOSIS. Broadest towards front part of shell, fairly evenly rounded from front to back, not tumid. Umbones not especially raised above dorsal part of shell. Carina rounded. Rapid increase in number of ribs towards the rear along line of greatest tumidity.

DESCRIPTION. Features of the genus are well shown in the material from the Redesdale Ironstone. The lunule and the escutcheon are distinct. A groove is present on either side of the carina outside of which there is an area between the groove and the rounded carina which has ribs less well developed than in front of the carina. The posterior has a

Fig. 9 *Myofossa (Myofossa) subarbitrata* (Dickins). Lower Permian, Nura Nura Member of Poole Sandstone Formation, 1·6 miles SW of Paradise Homestead, Canning Basin, Western Australia. Figs 9a–c, BMR CPC 3885, holotype; Fig. 9a, top view; Fig. 9b, oblique view showing siphonal area; Fig. 9c, view of left side; all ×3. Figs 9d–e, BMR CPC 3886, paratype A; Fig. 9d, view of left valve; Fig. 9e, top view; both ×2. Fig. 9f, BMR CPC 3887, paratype B, side view showing slight subumbonal sulcus, ×3.

significant gape. A narrow ligament groove is shown behind the umbo in BM L47314 and the hinge appears to be edentulous. The ribbing is distinctive. In the holotype, 16 ribs can be counted in the front part of the shell which are more or less concentric. The rib number increases rapidly by interpolation and 39 are present in the most tumid part of the shell; some of the added ribs make a distinct Y-like bifurcation with ribs at the front. A rounded posterior adductor muscle scar is shown in BM L3231 and the back part of the pallial line has a shallow sinus.

BM L47515, from Redesdale, has a well-preserved short nymph with a narrow, lunate ligament groove separating it from the escutcheon. Its inner surface is juxtaposed to the other nymph and does not project inwards from the remaining part of the posterior dorsal shell margin.

REMARKS. Hind (1900: 381) recognized that M'Coy (1851*a*: 174; 1855: 508; pl. 3f, figs 6, 6a, 7, 7a) had included two distinct shells in *Sanguinolites variabilis*. Elsewhere (p. 69) we have pointed out that the specimen in M'Coy, 1855: pl. 3f, fig. 6, 6a, is the lectotype of *Cosmomya variabilis* (M'Coy), selected by Hind (1900), and a new name is now required for the other specimen. Driscoll (1965: 91) included this species as a synonym of *Myofossa [Grammysia] omaliana* (de Koninck) in his redescription of that species. However, although *Myofossa hindi* has a similar shape to *M. omaliana*, the ornament, particularly the density and style of splitting of the ribs, is quite distinctive. Both species occur in the British Isles, but in rather different lithologies and apparently never together.

M. hindi is known only from the Carboniferous Limestone at Lowick and from the shell band in the Redesdale Ironstone, D_2 Zone, both in Northumberland. *M. omaliana*, on the other hand, occurs in Viséan limestone in Kildare and Limerick, commonly at Thorpe Cloud, Derbyshire, and in the lowest Namurian Great Limestone at Stanhope, Northumberland.

Myofossa (Myofossa) omaliana (de Koninck, 1842)
Figs 11a–d

1842   *Pholadomia omaliana* de Koninck: 65; pl. 5, fig. 4a–b.
1885   *Chaenomya omaliana* (de Koninck); de Koninck: 6; pl. 1, figs 12–15.
1900   *Sanguinolites omalianus* (de Koninck); Hind: 372–4; pl. 40, figs 17–24.

TYPE MATERIAL. Not seen, lent to de Koninck by M. Puys, from 'l'argile de Tournay', Tournaisian, Belgium.

REMARKS. This seems to be a very widespread species, occurring throughout the Lower Carboniferous and the base of the Upper Carboniferous, in Belgium, the British Isles and the United States. It has been extensively discussed by Driscoll (1965), but does not include *Sanguinolites variabilis* M'Coy, *pars* (=*Myofossa hindi* sp. nov.), see above.

Fig. 10 *Myofossa hindi* sp. nov. Lower Carboniferous, Viséan, Asbian, Redesdale Ironstone, Redesdale, Northumberland. Figs 10a–d, BM L47511, holotype, bivalved specimen (Hind, 1900: pl. 44, fig. 3); Fig. 10a, dorsal view; Fig. 10b, view of left side; Fig. 10c, ventral view; Fig. 10d, anterior view. Fig. 10e, BM L3231, paratype, view of left valve of steinkern; PA — posterior adductor, PS — pallial sinus. All ×2.

Myofossa (Myofossa) costellata (M'Coy, 1851)
Figs 1, 12a–d

1851a *Leptodomus costellatus* M'Coy: 174.

1855   *Leptodomus costellatus* M'Coy; M'Coy: 508; pl. 3f, fig. 5.

1900   *Sanguinolites costellatus* M'Coy; Hind: 377–9; pl. 41, figs 8–10.

SYNTYPES. SM E13273 (M'Coy, 1855: pl. 3f, fig. 5), and four unfigured syntypes, SM E9319–22, all from the Lower Limestone Series (Viséan) of Craige, Kilmarnock, Ayrshire, Scotland.

REMARKS ON NOMENCLATURE. *Myofossa costellata* is a junior secondary homonym of *Sanguinolites costellatus* M'Coy, 1844, considered by Hind (1900: 379) to be the posterior end of an internal cast of *Solemya costellata*.

DISCUSSION. *Myofossa costellata* is evenly ribbed, and has 50% more ribs per unit distance from the umbones than *M.*

hindi. Like *M. hindi*, it also occurs in very fine-grained facies. It is more elongate than the other species described here. It has not been found to occur with either *M. omaliana* or *M. hindi*. The fine ribbing recalls that of *Ragozinia*, but it dies not have the smoother sub-umbonal part of the flank.

Subgenus *RAGOZINIA* Muromzeva, 1984

TYPE SPECIES. *Myofossa (Ragozinia) dembskajae* (Muromzeva & Guskov, *in* Muromzeva 1984) (? = *Myofossa (Ragozinia) amatopensis* (Thomas, 1928)).

Myofossa (Ragozinia) amatopensis (Thomas, 1928)
Figs 13a–e

1928   ?*Sanguinolites amatopensis* Thomas: 224–5; pl. 6, figs 10, 10a.

1963   *Chaenomya* sp. Dickins: 51; pl. 8, figs 12–16.

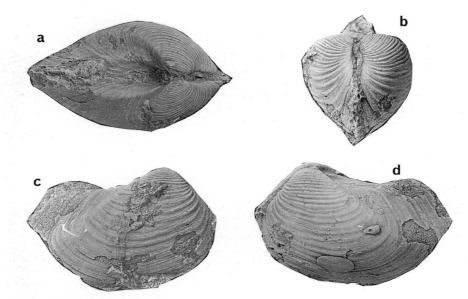

Fig. 11 *Myofossa omaliana* (de Koninck). Lower Carboniferous, Kildare, Ireland; BM 26327; Fig. 11a, top view; Fig. 11b, anterior view; Fig. 11c, view of right side; Fig. 11d, view of left side. All ×1.

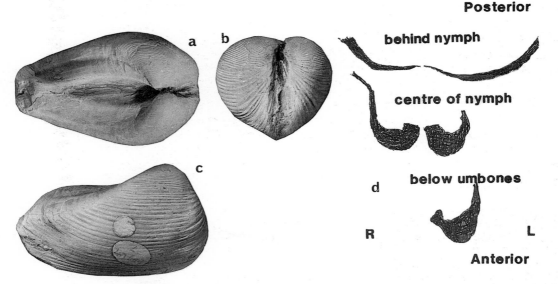

Fig. 12 *Myofossa costellata* (M'Coy). Lower Limestone Shale, Lower Carboniferous, Gurdy railway cutting, Gurdy, near Beith, Ayrshire, Scotland; BM L47489; Fig. 12a, top view; Fig. 12b, anterior view; Fig. 12c, view of right side; all ×1·25. Fig. 12d, BM L46425, transverse sections through ligament nymph, with detached nymph of right valve re-orientated to original position; approx. ×13; see also Fig. 1.

?1976   *Myonia (Myonia) gorskyi* Astafieva-Urbaitis: 32;
       pl. 3, fig. 5.
cf. 1984   *Ragozinia gorskyi* (Astafieva-Urbaitis); Muromzeva:
       113–14; pl. 41, figs 6–8.
?1984   *Ragozinia dembskajae* Muromzeva & Guskov, *in*
       Muromzeva: 114; pl. 41. figs 2, 4, 5.

HOLOTYPE. SM A4971, from the 'Goniatite Bed', Parinas Quebrada, Peru, Permian (not Upper Carboniferous as interpreted by Thomas); Barrington-Brown collection. The type material of this species was apparently mistakenly associated with Pennsylvanian ammonoids when collected.

OTHER MATERIAL. BM L9448, Irwin District, Perth Basin, Western Australia, no horizon recorded, but almost certainly Fossil Cliff Formation, Late Sakmarian; University of Western Australia, type no. 45374 (Dickins 1963: pl. 8, figs 12–13), from the Fossil Cliff Formation, Fossil Cliff, Perth Basin; Commonwealth Palaeontological Collection (C.P.C.) No. 3881, from the Callytharra Formation, Carnarvon Basin; C.P.C. No. 3882, from the Nura Nura Member of the Canning Basin. These are all thought to be of Lower Permian, Late Sakmarian age.

DIMENSIONS. BM L9498: length 84mm, height 48mm, width (two valves) 35mm, gape *c*. 9mm.

DISCUSSION. During reorganization of the collections at the British Museum (Natural History) BM L9498 was found with Mesozoic specimens. It was recognized as specifically identical

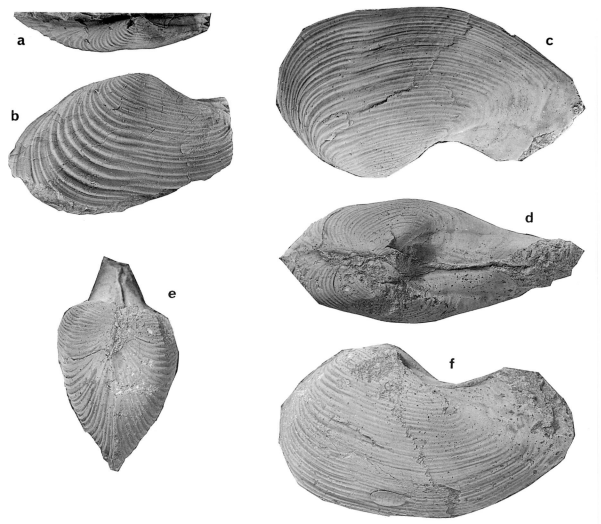

Fig. 13 *Myofossa (Ragozinia) amatopensis* (Thomas). Figs 13a–b, Permian, 'Goniatite' bed, Parinas Quebrada, Amatope Mountains, northwest Peru; SM A4971, Barrington Brown Collection, holotype; Fig. 13a, top view; Fig. 13b, left side. Figs 13c–f, Lower Permian, Sakmarian, Fossil Cliff Formation, Irwin District, Western Australia; BM L9498; Fig. 13c, right side; Fig. 13d, top view; Fig. 13e, anterior view; Fig. 13f, left side. All ×1.

with the specimens described by Dickins and there is now adequate material to indicate that this is a very widespread species. The matrix of the newly discovered Australian specimen, BM L9498, is that of the Fossil Cliff Formation. The label with the specimen reads 'Irwin District, Western Australia. Presented H. P. Woodward, May 1892'. The specimen apparently came from Fossil Cliff.

Although Runnegar (1969: 285) has suggested this species belongs to *Australomya*, the features of the escutcheon show that it belongs to *Myofossa*. When we compared the Australian material with Thomas' Peruvian holotype we were unable to find any significant differences. The Western Australian specimens and the holotype show considerable similarity to the Carboniferous species assigned in this paper to *Myofossa*, and especially to *S. (Ragozinia) gorskyi* from the Kungurian of the USSR, though the specimens from Western Australia are rather larger. Muromzeva (1984) distinguished a separate species, *M. (R). dembskajae*, which had a much more attenuated posterior. We have not been able to examine any

of her material, but we suspect the differences may be due to crushing. The posterior part of the shell, however, is more upturned in *M. (R.) amatopensis* than in the Carboniferous species, probably reflecting adaptation to deeper burrowing. *M. (R.) amatopensis* differs from its contemporary, *M. subarbitrata*, in lacking a concave anterior margin, its much finer ribbing and in being upturned at the back.

Myofossa sp. subgenus undetermined   Fig. 14

MATERIAL. USNM Ass 161469, Graham (top of Jacksboro), Old Gunter Road, 0.5 miles south of Texas 24, 5 miles NE of Jacksboro, Texas. USNM DR II 1506, Renfro Collection, loc. 1506, similar horizon, Upper Pennsylvanian, Texas.

DESCRIPTION. This apparently undescribed species has sculpture of the same general pattern as *Myofossa*, s. str., but much more exaggerated. There is an anterior field of prominent comarginal ribs with obvious radiating pustulose striae.

These striae are more randomly arranged within the ill-defined lunule, which is terminated at the sub-umbonal sulcus where it meets the ventral margin anterior to the umbones at a distinct sinus in the shell margins. The larger, main area of the flank is an area of low comarginal ribs and radial striae. These pustulose radial striae stop abruptly at the posterior area, which has strong transverse ridges parallel to the siphonal margins. There is no carina to the escutcheon which has periostracal creases on the shell surface and an ill-defined lunule.

REMARKS. *Myofossa* sp. has a wider central field of comarginal ribs, extending well anterior to the umbones, than ?*Grammysiopsis maria* (Worthen; Runnegar 1974: pl. 3, figs 1, 2, 11). The species may eventually prove to belong to *Grammysiopsis*.

Fig. 14 *Myofossa* sp. nov. Upper Pennsylvanian, Graham Formation (top of Jacksboro Member), Old Gunter Road, 0·5 mile south of Texas 24, 5 miles NE of Jacksboro, Texas; USNM Ass 161469; Fig. 14a, top view, with posterior end tilted upwards; Fig. 14b, left side; both ×1·7.

Genus *PALAEOCORBULA* Cowper Reed, 1932
Figs 15a–c

TYPE SPECIES. *Palaeocorbula difficilis* (Cowper Reed, 1932), by monotypy.

DISCUSSION. There are strong similarities between the shape and sculpture of *Myofossa* and *Palaeocorbula*. This latter genus is based on a single specimen of *P. difficilis* from Middle Horizon One, of the Lower Permian Agglomerate Slates of Kashmir. The specimen is housed in the collections of the Geological Survey of India, Calcutta; we have been kindly supplied with photographs of the type by S. C. Shah, Director of Palaeontology and Stratigraphy at the Geological Survey of India, which are reproduced here (Fig. 15). In our interpretation the short, coarse comarginal ribs are anterior and they bifurcate at a short distance from the anterior margin, in a similar manner to most other species of *Myofossa*. The posterior siphonal area is nearly smooth. In the holotype the valves are mutually displaced and much of the corselet of

the two valves is apparently missing (Fig. 15). It appears to us that the posterior of the left valve has been somewhat foreshortened by diagenetic or tectonic deformation. These vagaries of preservation apparently led Cowper Reed to interpret the animal as inequivalve and place the genus in the Corbulidae. It is the strong similarity in style of ribbing between *Palaeocorbula* and *Myofossa subarbitrata* (Fig. 9), and a number of other species belonging to the genus, that allows us to interpret *Palaeocorbula* as a close relative of *Myofossa*.

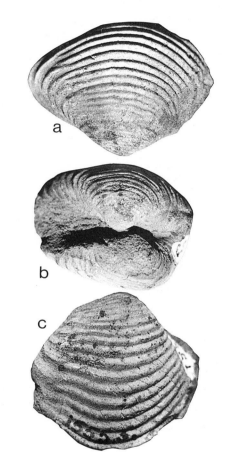

Fig. 15 *Palaeocorbula difficilis* (Cowper Reed). Upper Permian, Kashmir, India; GSI 15553, holotype; Fig. 15a, right side; Fig. 15b, top view; Fig. 15c, left side; all ×1·5.

Genus *COSMOMYA* Holdhaus, 1913

TYPE SPECIES. *Cosmomya egraria* Holdhaus (1913: 446; pl. 94, fig. 3a–c), by monotypy. A plaster cast of the type specimen in the GSI has been figured by Dickins & Shah (1965: pl. 17, fig. 13–14).

SYNONYMY. *Palaeocosmomya* Fletcher, 1946 (type species, *P. teicherti* Fletcher, 1946 by original designation (?=*Cosmomya egraria* Holdhaus, 1913; ? = *Cosmomya artiensis* (Krotova, 1885: 255; pl. 3, fig. 20)).

DISCUSSION. The relationship between *Cosmomya* and *Palaeocosmomya* has been discussed by Dickins & Shah (1965). At that time the family relationships of the genus were obscure but the early species from the British Lower

Carboniferous, particularly the material in the British Museum (Natural History) described by Hind (1900) as *Sanguinolites v-scriptus*, show a transition between early species of *Myofossa* to *Cosmomya* by the gradual acquisition of typical eccentric zig-zagging ribs. This has led us to place *Cosmomya* in the Sanguinolitidae. *Cosmomya* differs from *Grammysiopsis* Chernychev, 1950 and *Pentagrammysia* Chernychev, 1950 by the position of these ribs, and it differs from *Praeundulomya* Dickins, 1957, by the shell shape and by lacking the posterior elongate shell thickenings that run close to the hinge in that genus. *Pentagrammysia*, a genus that has developed the ribbing style of the Mesozoic genus *Goniomya* independently, seems to have evolved in central or eastern Asia separately from *Cosmomya*, but we suggest that it also has common ancestry with a species of *Myofossa*. Several species of *Pentagrammysia* are illustrated in Muromzeva's publications (particularly 1974: pls 21–23). In the British species described below, ascending stratigraphical position reflects increasing complexity of ribbing pattern. Early species of the Mesozoic genus *Goniomya* differ from the type species of *Cosmomya*, *Pentagrammysia* and *Siphogrammysia* in having a sub-umbonal V in the ribbing which slopes downwards and backwards, and is horizontally truncated at least in the umbonal area. We have not been able to decide whether *Goniomya* is directly descended from one of these Upper Palaeozoic genera with V ribbing, or whether *Goniomya* has developed this style of ribbing independently. The material figured by Runnegar (1974: pl. 3, figs 1, 2, 11) as *Cosmomya maria* (Worthen) is better placed in *Grammysiopsis*, and if both fragments do in fact belong to the same species, they do not differ significantly from *G. bisulcatiformis* Muromzeva & Kagarmanov (*in* Muromzeva 1974: 108; pl. 23, figs 21–22).

Cosmomya v-scripta (Hind, 1900)   Figs 16a–b

1900   *Sanguinolites v-scriptus* Hind: 382; pl. 42, figs 6, 6a, 7, 7a only.

LECTOTYPE. BM L47495 (BM L46533 is the counterpart), here designated, is Hind's (1900: pl. 42, figs 7, 7a) figured specimen; it is from a marine sand approximately 500ft. below the third bed of Millstone Grit, probably E2 Zone, Congleton Edge, Cheshire, England. Fig. 16.

PARALECTOTYPES. BM L47494, the specimen figured by Hind (1900: pl. 42, figs 6, 6a), the same horizon and locality as the lectotype. The second paralectotype, BM L47493 also figured by Hind (1900: pl. 42, figs 5, 5a), should be referred to *Myofossa hindi*; it is from the Redesdale Ironstone. We have found no other material.

DIAGNOSIS. Escutcheon and groove inside rounded posterior carina developed as in *Myofossa*. Posterior V of ribbing distinct, but only a slight anterior V. A slight sulcus runs from the umbo to the anterior part of the dorsal margin in the position of the slight anterior V of the ribbing.

DESCRIPTION. Little can be added to Hind's description and to the diagnosis. The three specimens are deformed in different directions, and measurements are not meaningful. The lectotype is compressed laterally, and the paralectotype BM L47494 is elongated front to back. Despite this the principal characteristics of the species seem clear enough. The prominent umbo of BM L47493 from the Redesdale Ironstone is apparently an artefact, and although this specimen is quite small, it is clearly a different species.

DISCUSSION. The relationship of this species to *Myofossa* on the one hand and to the type species of *Cosmomya* on the other hand seems well established. The characters associated with the escutcheon are essentially those of *Myofossa*. The anterior sulcus and ribbing represent the basic pattern of *Cosmomya*. The anterior V in the ribbing, however, is only slightly developed and the posterior V is less distinct. The pattern, however, is so close that there is little doubt that *Cosmomya v-scripta* is an ancestral species of the genus.

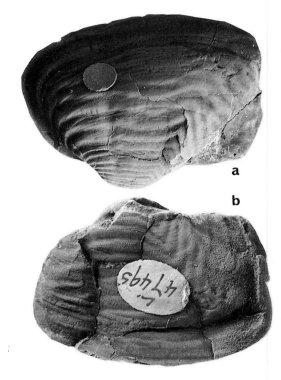

Fig. 16   *Cosmomya v-scripta* (Hind). Namurian (probably E2), marine sand *c.* 500 ft below third Millstone Grit, Congleton Edge, Cheshire; BM L47495, **lectotype** (selected herein); Fig. 16a, right valve; Fig. 16b, left valve; both ×2·5.

C. v-scripta resembles '*Grammysiopsis*' *omolonicus* Muromzeva (1974: 108; pl. 21, figs 1–3), which comes from near the Permo-Carboniferous boundary in the Omolonsk Massif, Irbichan, U.S.S.R. It does not, however, have such an extensive posterior gape and area. '*G.*' *bisulcatiformis* Muromzeva & Kazamanov (*in* Muromzeva 1974: 108; pl. 22, fig. 21) is difficult to distinguish from '*G.*' *omolonicus*; both species differ in having a considerably deeper V in the ribbing.

Cosmomya variabilis (M'Coy, 1851)   Figs 17a–d

1851a   *Sanguinolites variabilis* M'Coy: 174, *pars*.
1854   *Sanguinolites variabilis* Morris: 223.
1855   *Sanguinolites variabilis* M'Coy: 508; pl. 3f, figs 6, 6a only.
1900   *Allorisma variabilis* (M'Coy); Hind: 424; pl. 44, fig. 2.
1900   *Sanguinolites interruptus* Hind: 383; pl. 42, figs 8, 9, 10; pl. 49, fig. 10.

Fig. 17 *Cosmomya variabilis* (M'Coy). Figs 17a–b, High Viséan, Lowick, Northumberland; SM E1100, lectotype; Fig. 17a, angled top view; Fig. 17b, side view (note that the anterior ventral part of the shell is missing). Fig. 17c, Lowest Namurian, Main Limestone, Stanhope, Weardale, Northumberland; BM PL1598, left valve. Fig. 17d, Lower Carboniferous, Viséan, Thorpe Cloud; BM L47497, syntype of *Sanguinolites interruptus* Hind, left side. All ×1·5.

LECTOTYPE. SM E1100 (M'Coy, 1855: pl. 3f, figs 6, 6a), subsequently designated by Hind (1900: description of pl. 44, fig. 2).

SYNTYPES OF *Sanguinolites interruptus*. These are BM L47496 (Hind, 1900: pl. 42, fig. 8), BM L47497 (Hind, 1900: pl. 42, fig. 10), BM L47538 (Hind, 1900: pl. 49, fig. 10), and one specimen from the same locality said by Hind to be in the collection of Mr Holroyd of Manchester; all are from the Lower Carboniferous, Viséan of Thorpe Cloud, Derbyshire, England.

OTHER MATERIAL. BM L46434–40, 7 specimens from Castleton or Thorpe Cloud in the Hind Collection; BM L43637, from Narrowdale, and BM L46433 from Gateham, between Wetton Hall and Narrowdale; all Lower Carboniferous, Viséan.

DESCRIPTION. Oval, tumid shell with distinct rugae, which are straight or slightly curved and comarginal in the mid part of the flank, but diverge across the growth lines in the anterior part forming a slight V. There is an indistinct V in the posterior part before they curve round with the siphonal margin. The greatest tumidity is posterior to the umbones. Umbo moderately distinct, pointed towards the front. The lunule is obscure and the escutcheon is typical for *Cosmomya* and *Myofossa*. There is a rounded carina ventrally, and above this a groove running to the posterior margin of the posterior flattened area; above this the escutcheon proper is an elongated flat area bordering the external margin of the shell. From comparison with similar shells, a ligament groove and nymph might be expected at the anterior part of this flat area. A slight posterior gape was probably present. A shallow sulcus runs from the umbo to the ventral margin.

DIMENSIONS (mm):	Length	Height	Width
BM L47497, LV	39	23	9
BM L47496, LV	31	19	8

DISCUSSION. Two distinct species are represented in M'Coy's (1855) illustrations: i.e. his pl. 3f, figs 6 and 6a, and his pl. 3f, figs 7 and 7a. The identity of his third specimen (pl. 3f, fig. 8) is not clear, but it may be the same as his pl. 3f, figs 6 and 6a, as suggested by M'Coy himself when he described it as an oblong variety.

Hind (1900: 381) recognized these two different species and he referred them to two genera, *Sanguinolites* and *Allorisma*, both with the specific name *variabilis* attributed to M'Coy. Hind's choice of lectotype is in accordance with I.C.Z.N. Article 74 (a) (v), which states that a specimen that was not originally described as a syntype, i.e. was described as a variety, is not available for choice as lectotype. He incorrectly determined the specimen as an *Allorisma* (*Wilkingia* in terms of present usage), apparently failing to notice the incipient Vs of the discordant ribbing present on the lectotype; he included specimens of *Wilkingia regularis* de Verneuil in the same species. The lectotype, although damaged in the anterior region, is clearly the same species as BM PL1598, from the Lower Namurian Main Limestone of Stanhope near Durham, see Fig. 17c.

The V in the ribbing of this species is the only apparent difference from the genus *Myofossa*, but we consider this sufficient to allow us to recognize it as an early species of *Cosmomya*. It differs from *Cosmomya v-scripta* in having a less well marked posterior V in its ribbing. BM L46433 has more steeply dipping anterior transcurrent ribs than the other specimens and is in this way similar to the anterior flank of *Pentagrammysia*.

Genus *SIPHOGRAMMYSIA* Chernychev, 1950

TYPE SPECIES. *Pholadomya kasanensis* Geinitz 1880. Permian. ?Kazanian, Kazan, U.S.S.R.

DISCUSSION. *Siphogrammysia* resembles some other sanguinolitids in shape and has divergent ribs forming Vs in a similar way to *Cosmomya* and *Pentagrammysia* (Chernychev 1950; ?=*Manankovia* Astafieva-Urbaitis, 1983), but the discordant ribs are much bolder in design and more prominent. They are convergent in pattern with ribs of the trigoniacean, *Iotrigonia*, and a number of Unionacea including *Trigonioides*. A number of *Siphogrammysia* species have been described from the Upper Permian of Kazan and the Taimyr Peninsula. Here (Fig. 18) we illustrate some material from Malaysia which may be the same age.

Siphogrammysia cf. *kasanensis* (Geinitz, 1880)   Fig. 18

cf. 1880   *Pholadomya kasanensis* Geinitz: 38–9; pl. 6, figs 23, 23a.
cf. 1894   ?*Goniomya kasanensis* (Geinitz); Nechaev: 314–16; pl. 10, figs 22, 26.
     1950   *Siphogrammysia kasanensis* (Geinitz); Chernychev: 26–7; pl. 7, figs 61–3.

MATERIAL. BM PL5001 and BM PL5009, from Geological Survey of Malaysia locs. 106/RF/15 and 116/F/9; Labis area of Johore, Malaysia; in weathered silty shales associated with poorly preserved Ammonoidea, possibly mid-Permian *Agathiceras* sp.

Fig. 18 *Siphogrammysia* cf. *kasanensis* (Geinitz). ?Upper Permian, Kazanian, Malaysia, locality 116/F/9B; BM PL5001; ×2·4.

DISCUSSION. The anterior flank of this species has prominent, broad transcurrent ribs which 'V' sharply, and near-vertical less prominent ribs below the umbones. A second V delimits the siphonal margin at its ventral point, with the V opening to the posterior. Yet a third V runs close to the dorsal part of the dorsal or siphonal area, again open to the posterior. This species differs from *Sanguinolites inordinata* Thomas (1928:

226–7; pl. 6, figs 8, 8b; pl. 8, fig. 7), which appears to be conspecific with *Goniomya kasanensis*, as interpreted by Lutkevich & Lobanova (1960:, 86; pl. 11, figs 1, 2), in having no break in the transcurrent ribs of the anterior flank. The latter two species clearly belong to *Cosmomya* rather than to *Siphogrammysia*, and are probably conspecific with *C. egraria* Holdhaus.

Genus *GILBERTSONIA* nov.

TYPE SPECIES. *Sanguinolaria gibbosa* J. de C. Sowerby, 1827, here designated.

DESCRIPTION. Nearly smooth, with comarginal growth lines and obscure comarginal ribs. No surface pustules observed. Umbones forward of mid-point, rounded and opisthogyral. A wide posterior dorsal area between the umbones and the siphonal margins joins the flank with a gentle change in shell slope, not defined by any feature of the ornament. Shell thin, inflated and elongate; striations on the inner shell surface possibly represent migrating points of mantle attachment. Anterior margin rounded and protruding; ventral margin sinuous, following a shallow, near vertical ventrolateral sulcus, sub-parallel to the hinge. Lunule distinct but not carinate, smooth; escutcheon long, narrow and carinate. The adductors are apparently sub-equal although the anterior one is not well preserved on any specimen we have examined. The posterior adductor is large, sub-rounded and spans the wide posterior dorsal area. The pallial line is without a sinus; it passes ventrally and posteriorly from the lower posterior edge of the posterior adductor parallel to the sloping posterior shell margins. The hinge plate is moderately thick, with a moderately long, very narrow, barely protruding nymph. The dorsal margins are opposed (adpressed) in a straight line well to the posterior. This indicates that they were joined by periostracal ligament.

OTHER SPECIES. *Unio ansticei* J. de C. Sowerby (1840: pl. 39), and two apparently unnamed species from the Upper Carboniferous, Fort Jackson area, Texas.

REMARKS. *Gilbertsonia* resembles *Pachymya*, but that Mesozoic genus does not have a clearly defined lunule and has a much more robust ligament nymph and particularly prominent lines of shell surface pustules. Species of *Pachymya* are usually more thick-shelled than *Gilbertsonia*. The position of the posterior part of the pallial line also differs, and *Pachymya* has a shallow but distinct pallial sinus.

   Eopleurophorus [*Sanguinolites*] *hibernicus* Hind has a very similar shell shape but a less sinuous ventral margin; it also has a posterior dorsal area with low ribs.

Gilbertsonia gibbosa (J. de C. Sowerby, 1827)
                                        Figs 19a–g

1827   *Sanguinolaria gibbosa* J. de C. Sowerby: **6**: 92; pl. 548, fig. 3.
1836   *Sanguinolaria tumida* Phillips: 209; pl. V, fig. 3.
1844   *Allorisma gibbosa* (J. de C. Sowerby); King: 315.
?1844   *Sanguinolites contortus* M'Coy: pl. 19, fig. 3.
?1885   *Sanguinolites luxurians* de Koninck: 73; pl. 16, figs 1–3.
?1885   *Sanguinolites tumidus* (Phillips); de Koninck: 81; pl. 16, fig. 6.

Fig. 19 *Gilbertsonia gibbosa* (J. de C. Sowerby). Lower Carboniferous, Viséan. Figs 19a–c, 'Queen's County' (= County Laois), Ireland; BM 43056, holotype;) Fig. 19a, top; Fig. 19b, left side; Fig. 19c, bottom; all ×1·1. Figs 19d–e, Kildare, Ireland; BM L231, Tennant Collection; Fig. 19d, right side; Fig. 19e, top; both ×1. Fig. 19f, ?Bolland, Yorkshire; BM 97200, Gilbertson Collection, slightly oblique view of anterior of left valve showing anterior adductor; × 1·5. Fig. 19g, Kildare, Ireland, BM L47930, right side, with posterior adductor PA, and pallial line without pallial sinus PL.

?1885   *Sanguinolites portlocki* de Koninck: 82; pl. 16, fig. 11.
?1885   *Chaenomya requiana* (de Rychholdt); de Koninck: 7; pl. 1, fig. 11.
?1900   *Sanguinolites luxurians* de Koninck; Hind: 402–4; pl. 46, figs 3–5.

HOLOTYPE. BM 43056 is the figured and only known type specimen; from the Lower Carboniferous of 'Queen's County' (= County Laois), Ireland.

TYPES OF *Sanguinolaria tumida* PHILLIPS. The specimen figured by Phillips (1836: pl. V, fig. 3) has not been discovered in the Phillips collection at Oxford nor in the Gilbertson collection at the BM(NH). Phillips said his figure was reduced from a large Irish specimen. Specimens from the Gilbertson collection listed as *S. tumida* in the Gilbertson Catalogue, but not labelled as such by Phillips and not belonging to the species as here understood (that is, not belonging to the same species as illustrated by Phillips), are not accepted here as syntypes. These include BM 97164 belonging to the species here recognized as *Myofossa omaliana* de Koninck (p. 63).

OTHER MATERIAL:
BM L231 from the Carboniferous Limestone of Kildare, Ireland; Tennant Collection, Fig. 19d–e (Hind identified this specimen as *S. luxurians*).
BM 97187 (41.6.7.132 in Gray Catalogue; 92g in Gilbertson Catalogue; *Isocardia* sp.; Phillips MS: pl. 5b, fig. 34; see below).
BM 36937, Carboniferous Limestone, Clane, Kildare, Ireland, Pratt Collection.

BM L13486, Carboniferous Limestone, Derbyshire, England.

BM L47930, Lower Carboniferous, Kildare, no other details.

BM L45242, Lower Carboniferous, St. Doolaghs, Co. Dublin, Ireland (labelled *Allorisma ansticei* Sowerby); Hind Collection.

BM(NH). Bancroft Collection no. 786, Clane, Kildare, Ireland.

REMARKS. Phillips' species *Sanguinolaria tumida* has to be interpreted from his figure (Phillips, 1836: pl. 5, fig. 3). Although we have been able to examine the mock-ups for the original plate (with kind help from Mr P. Powell of the Oxford University Museum) we have not been able to locate a type specimen, either in Phillips' own collection at the Oxford University Museum, or in the Gilbertson Collection at the BM(NH). The original manuscript and published figures seem to portray a species which may be considered to be a synonym of *Gilbertsonia gibbosa*. Unfortunately, the specimens in the Gilbertson Collection which we attribute to this species are not named as such in Gilbertson's manuscript catalogue, housed in the Palaeontology Library at the BM (NH). A further difficulty is that a specimen labelled no. 85 in the Gilbertson Catalogue, and there identified as *S. tumida*, is clearly not the specimen or species figured by Phillips; it is a *Myofossa omaliana* (p. 63). Hind apparently thought that the specimen BM 97200 was Phillips' figured specimen of *S. tumida*, but it does not have any individual features in common with the figure. We are therefore unable to identify any type material of *Sanguinolaria tumida* Phillips. Because we consider that species to be a junior subjective synonym of *Sanguinolaria gibbosa* J. de C. Sowerby, by identification of Phillips' figure with that species, we do not think it would be advantageous to create a neotype for Phillips' species. It is possible that *Sanguinolites luxurians* de Koninck, from the slightly older Calcschiste de Tournai, is the same species; but in de Koninck's figure the umbones are closer to the anterior. *S. luxurians* seems to have been based on only one specimen. De Koninck grouped his *S. luxurians* with species that have two diagonal folds (ribs?); if such ribs are present at all in *G. gibbosa* they are very indistinct. They are not at all comparable with the radiating sculpture that occurs on the corselet in some *Sanguinolites* and Permophoridae.

Sanguinolites contortus M'Coy may be a distorted specimen of the present species; we have not been able to examine the holotype. At present we treat *S. contortus* as a *nomen dubium*.

Subfamily UNDULOMYINAE Astafieva-Urbaitis, 1984

In this subfamily are placed *Wilkingia*, *Praeundulomya*, *Undulomya*, *Exechorhynchus*, *Dyasmya* gen. nov. and probably *Manankovia*. The subfamily contains elongate forms progressively adapted for deep burrowing. A lunule and escutcheon are present, with the ligament lodged in a groove at the front end of the escutcheonal area on a narrow ligament nymph. The external shell surface is granular, except for the escutcheon, with aligned periostracal spicules. Species of *Wilkingia* are known to have a deep pallial sinus. Runnegar (1969: 287, fig. 53c) records a pallial sinus in a specimen from the topmost Carboniferous or lowermost Permian of the U.S.A. The other characters of *Praeundulomya* and *Undulomya* link them to *Wilkingia*, from which we infer that they may also have had a deep pallial sinus. Just

noticeable in *Wilkingia regularis*, but better developed in *W. maxima*, is a low, rounded, elongate rib present on the internal shell surface, running from the umbones towards the posterior margin. In *Praeundulomya* and *Undulomya* a second rib occurs, a little lower, which reaches the posterior margin at a point where the division between the two siphons would be expected. This feature also occurs in *Siliquimya*, ?Permophoridae.

The genera of this subfamily do not have the deeply inset anterior adductor, with a shell thickening immediately behind it, characteristic of *Siliquimya*. *Undulomya* has prominent V-shaped ribs, very similar to those in *Pentagrammysia* Chernychev, 1950, but that Carboniferous genus does not have the posterior interior radiating ribs that link it with *Praeundulomya* and *Wilkingia*. We therefore conclude that the V ribs have developed independently in this case. Astafieva-Urbaitis (1974b: fig. 1) illustrates a species from Kazakhstan with transcurrent ribs on the anterior only, intermediate between *Undulomya* and *Praeundulomya*. The Undulomyinae may be distinguished from *Vacunella* Waterhouse 1965 by the form of the escutcheon; in *Vacunella* the escutcheon does not have a sharp carinate edge and it is ill-defined distally where the dorsal margins pass evenly into the posterior siphonal gape. *Vacunella* also has much broader nymphs. *Myonia* Dana 1847, here also included in the Vacunellinae, has no pallial sinus.

Genus *UNDULOMYA* Fletcher, 1946

TYPE SPECIES. *U. pleiopleura* Fletcher, by original designation, = *Goniomya singaporensis* Newton, 1906.

Undulomya singaporensis Newton, 1906

?1906   *Goniomya scrivenori* Newton, 49: pl. 25, fig. 1.
  1906   *Goniomya singaporensis* Newton: 493; pl. 25, figs 2–3.
  1913   *Goniomya uhligi* Holdhaus: 450; pl. 94, fig. 2.
?1928   *Sanguinolites deportatus* Thomas: 229; pl. 6, figs 6.
?1928   *Sanguinolites insolitus* Thomas: 228–9; pl. 6, figs 11–12.
  1946   *Undulomya pleiopleura* Fletcher: 399–400; pl. 34, figs 1–5; pl. 35, fig. 1.
  1956   *Undulomya pleiopleura* Fletcher; Dickins: 29; pl. 4, figs 6–8.

TYPE MATERIAL. The two syntypes of *Goniomya singaporensis* are BM L19154, from the Permian, possibly Artinskian, Singapore (J. B. Scrivenor collection, *ex* Mr Guthrie), mistakenly described by Newton as Middle Jurassic; and BM L19173, part and counterpart of a fragment (Mr Hanitch collection). BM L19153 (apparently missing) is the holotype of *Goniomya scrivenori*. The holotype of *Goniomya uhligi*, from a geode in black shales below the Werfen Beds, NW of Kunplong, SW of the Niti Pass (horizon mistakenly doubted by Holdhaus (1913) and changed to Spiti Shales), is in the museum of the Geological Survey of India, Calcutta, and is from the same locality as *Cosmomya egraria* Holdhaus 1913. The type material of *Sanguinolites deportatus* and *S. insolitus*, from the 'Goniatite Bed' and Steel Hill, Parinas Quebrada, Amatope Mountains, Peru, is in the Sedgwick Museum, Cambridge; it is probably Lower Permian in age, but is apparently associated in the collections with Pennsylvanian ammonoids.

REMARKS. This species has been well described and illustrated by Fletcher (1946), but examination of material from Singapore and Peru convinces us that it has a wide geographical range outside Australia and that there are a number of older names available for it. While the species has almost exactly the same shape a *Praeundulomya maxima*, with identical escutcheon and posterior internal ribs, it has developed a very pronounced V pattern of ribs with the line bisecting the angle of V sloping slightly backwards below the umbones. It is the similarities, coupled with what appears to be a good intermediate discovered by one of us (Astafieva-Urbaitis 1974*b*), which lead us to suggest the close relationship between the two genera. We have been unable to discover the posterior part of the pallial line in any of the material we have examined, but assume that the species would have been siphonate with a deep pallial sinus because that feature is present in the presumed ancestor and other members of the subfamily.

Genus *WILKINGIA* Wilson, 1959

TYPE SPECIES. *Venus elliptica* Phillips, 1836 (*non V. elliptica* Lamarck 1818), by original designation (as interpreted by Wilson, 1959; = *Allorisma regularis* King, *in* de Verneuil, 1845).

SYNONYMS. *Allorismiella* Astafieva-Urbaitis, 1962: 36 (type species by original designation, *Allorisma sulcata* Hind (1900: 42; pl. 48, figs 3–11), wrongly quoted as '*Allorisma sulcata* Hind, 1896'; this is not *Hiatella sulcata* Fleming, 1828 (see p. 57) and not *Allorisma sulcata* King, 1844: 316. We consider *Allorisma sulcata* as interpreted by Hind and Astafieva-Urbaitis to be a junior subjective synonym of *Allorisma regularis* King, *in* de Verneuil 1845 (see below), and a junior objective synonym of *Venus elliptica* Phillips, *non* Lamarck.

Dulunomya Astafieva-Urbatis & Dickins, 1984; type species by original designation *Dulunomya serpukhovensis* Astafieva-Urbatis & Dickins. We consider *Dulunomya serpokhovensis* to be a junior subjective synonym of *Allorisma regularis* King, *in* de Verneuil 1845.

REMARKS. In order to establish the identity of the nominal subgenera *Wilkingia* and *Allorisma* we have had to overcome a series of compounded errors and misinterpretations. King (1850: 196–9; pls 16, 20) clearly distinguished the characters of these taxa but unfortunately, by reference to misidentified type species, applied the names in reverse. From his footnote 1 on p. 196, however, it is clear that in his first publication (King 1844) he used the name *Allorisma* both for members of the family Edmondiidae and for forms with a deep pallial sinus here included in the Undulomyinae. Indeed the internal rib below the hinge is described as occurring in *Allorisma sulcata* (Fleming) (King 1844: 316). In 1850, King in effect changed his mind over the identification of *Allorisma* when he discovered that the characters of *Sanguinolaria sulcata* of both Fleming and Phillips resembled those of *Edmondia*, and differed from the siphonate forms here included in the Undulomyinae. His intention was to use *Allorisma* for the siphonate species (King 1850: pl. 20, fig. 5 only), but unfortunately he misidentified his material with the non-siphonate species *Allorisma sulcata* (Fleming, 1828) which he

wrongly regarded as different from *Sanguinolaria sulcata* Phillips 1836 (see p. 58).

DIAGNOSIS. Lunule present and distinct escutcheon made up of an elongated flat area. Evenly rounded from front to back. Ribs constant in number. Pallial sinus deep.

DISCUSSION. In the Viséan and Namurian rocks, from the Moscow Basin to England and the U.S.A., there is a close knit group of species, which may only be distinguished in well-preserved specimens when subtle differences of shell shape and the form of the pallial sinus can be observed. These make up the genera *Wilkingia*, and *Praeundulomya* as here recognized. *Wilkingia* is of moderate size with sinuous ventral margins, and this group includes *W. regularis* and *W. ?transversa*. Species of *Praeundulomya* are large and include *P. maxima*, which has sub-parallel ventral and dorsal margins. The Permian *P. concentrica* also has this shape but is not so large.

It may prove in the future better to join *Wilkingia* with *Praeundulomya* as a junior subjective synonym, but at present we retain the two genera separated by the characters mentioned here.

Wilkingia regularis (King, *in* de Verneuil 1845)

Figs 20–24

?1836   *Venus elliptica* Phillips: pl. 2, fig. 7 (*non* Lamarck 1818).
 1845   *Allorisma regularis* King, *in* de Verneuil: 298; pl. 19, fig. 6 only.
 1850   *Allorisma sulcata* (Fleming); King: pl. 20, fig. 5.
 1900   *Allorisma sulcata* (Fleming); Hind: 320, 422–4; pl. 48, figs 3, 5, 6, 9–11.
?1900   *Allorisma sulcata* (Fleming); Hind: pl. 48, fig. 8 only.
 1900   *Allorisma variabilis* (M'Coy); Hind: pl. 48, figs, 1, 2 only.
?1950   *Tellinomorpha sarytschevae* Chernychev: 42; pl. 11, fig. 90.
?1950   ?*Tellinomorpha* sp. Chernychev: 43, fig. 92.
 1959   *Wilkingia elliptica* (Phillips); Wilson: 402–4; pl. 71, figs 1, 3–6.
?1959   *Wilkingia elliptica* (Phillips); Wilson: pl. 71, fig. 2 only.
 1962   *Allorismiella sulcata* (Hind) Astafieva-Urbaitis: 36.
?1962   *Allorismiella sulcata* (Hind); Astafieva-Urbaitis: 36–7, fig. 2.
 1962   *Allorismiella regulariformis* Astafieva-Urbaitis: 39–40, fig. 2.
 1984   *Dulunomya serpukhovensis* Astafieva-Urbaitis & Dickins: 38–9; pl. 2, figs 1–3.

TYPES. There are 12 syntypes in the de Verneuil Collection of the École des Mines (stored at present at the University of Lyon). Four of them are associated on a board bearing the register number 1743, and labelled '*Allorisma regularis* King, Sloboda, gouv. de Toula, Carbonifère. Coll. de Verneuil'; one of these is here selected lectotype (Fig. 20a–d), and the other three become paralectotypes. Other paralectotypes with similar locality labels are two specimens numbered 1744 and three specimens numbered 1745. Three more paralectotypes are on blocks numbered 1742 (two specimens; one of them was figured by de Verneuil (1845: pl. 21, fig. 11)) and 1746, labelled '*A. regularis* King, Valdai, Coll. de Verneuil'; we identify these three paralectotypes with the genus

Fig. 20 *Wilkingia regularis* (King, *in* de Verneuil). Lower Carboniferous, Viséan; Sloboda, Gouv. de Toula, Russia. Figs 20a–d, EMP 1743 (1 of 4), de Verneuil Collection, **lectotype** (selected herein), a silicified steinkern; Fig. 20a, right side; Fig. 20b, left side; Fig. 20c, anterior; Fig. 20d, dorsum. Figs 20e–f, EMP 1744, paralectotypes, views of right sides. Figs 20g–i, BM L18, ? *ex* de Verneuil Collection, ?paralectotype; Fig. 20g, left side; Fig. 20h, dorsum; Fig. 20i, latex cast of hinge area seen from the inside to show inner surface of nymphs. All ×0·88, except Fig. 20i, ×1·5.

Allorisma King. Finally, two specimens (BM L18) in the BM (NH) that have the collectors' number 50, in similar handwriting to one of the Valdai specimens, may also be original syntypes that now become paralectotypes.

SYNONYMS. Although under normal circumstances we consider it inadvisable to designate types for junior synonyms or other invalid names, we feel that it is necessary to do this in the present case in order to preserve the stability of the generic name *Wilkingia*.

1. *Venus elliptica* Phillips, 1836 (*non* Lamarck, 1818). **Neotype**, here designated, is BM L47526, the specimen figured by Hind (1900: pl. 48, fig. 4). Wilson (1959: 402) designated the figure in Phillips (1836: pl. 2, fig. 7) as the lectotype of this species. In the same publication, however, he stated that the original specimen could not be found in the collections of the Yorkshire Philosophical Society (at the Yorkshire Museum, York) where it might be expected to be stored, nor in the Hancock Museum, Newcastle-upon-Tyne, nor in the University College Galway, Ireland. We have also been unsuccessful in finding the specimen in the Gilbertson Collection in the BM(NH) (it is not listed in Gilbertson's manuscript catalogue). We have also searched in vain in Leeds City Museum and in the Phillips Collection at Oxford University Museum. In a series of Phillips' original drawings and mock-up plates for the *Geology of the Yorkshire Coast* kindly made available to us by Mr P. Powell of the Oxford University Museum, an original drawing of *Venus elliptica* is preserved together with the information that this specimen came from Harelaw, Northumberland and was in the collection of the Rev. C. V. Harcourt. These Harcourt specimens should be in the York Museum, and we would like to thank Dr Pyrrah

of that Museum for carrying out a further unsuccessful search.

Unfortunately it is very difficult to determine the taxonomic position of Phillips' lost specimen from his figure. Its characters include the elliptical shape with a rounded, non-sinuous venter, small size, indication of a lunule and broad, regular, distinct comarginal rugae. If indeed it was an anomalodesmatid it could be a synonym of one of four taxa, *Sanguinolaria sulcata* Fleming, *Sanguinolaria maxima* Portlock, *Allorisma regularis* King or *Pholadomya omaliana* de Koninck. *Siliquimya plicata* (Portlock), even ẗ this small size, has a distinctly more elongate shape and a sharper increase in curvature at the posterior ventral margin. In 1845, King (*in* de Verneuil) identified Phillips' species with one from Russia and indicated that it occurred in northern England. He used the name *Cardiomorpha sulcata* de Koninck 1842 for this species. In order to avoid further difficulty, the neotype chosen here makes *Venus elliptica* Phillips 1836 (*non* Lamarck, 1818) a subjective synonym of *Wilkingia regularis* (King, *in* de Verneuil 1845).

2. *Allorismiella sulcata* Astafieva-Urbaitis, 1962. In her original description of *Allorismiella*, Astafieva-Urbaitis designated *Allorisma sulcata* Hind as type species. Under ICZN Article 70c this must be construed as a deliberate misapplication of the name which Hind (1900) himself correctly attributed to Fleming 1828, but then misidentified. Following the provisions of this Article, the type species fixed by that action is deemed to be a new nominal species. In her original description, Astafieva-Urbaitis (1962: 40) referred to the specimens figured by Hind (1900: pl. 48 [pl. 18 cited in error], figs 3–11), the specimen figured by Fedotov (1932: pl. 10, fig. 5), and other specimens she had

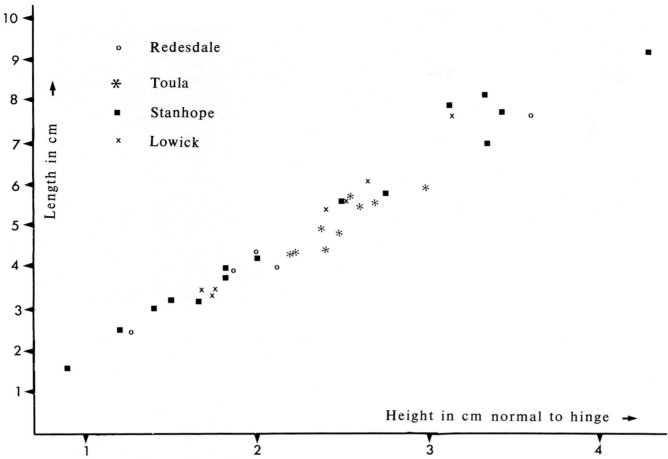

Fig. 21 *Wilkingia regularis* (King, *in* de Verneuil). Scatter diagram of height/length ratios of the four samples: (1) Redesdale Ironstone (Viséan) and (2) Main Limestone (Lower Namurian), both from Stanhope, Weardale; (3) Upper Viséan or Lowest Namurian, Lowick; (4) the type series from Sloboda (Viséan) (excluding specimens of *Allorisma*); this shows that the Redesdale material does not have a different height/length ratio from the other samples.

in front of her; all must be considered to be the type series of her new species. BM L47526, the specimen figured by Hind (1900: pl. 48, fig. 4), is here designated lectotype. By this action BM L47526 becomes the type specimen of both *Venus elliptica* Phillips (*non* Lamarck) and *Allorismiella sulcata* Astafieva-Urbaitis, so they are objective synonyms. Consequently the genera *Wilkingia* and *Allorismiella* become objective synonyms. In our opinion the valid name for the type species of *Wilkingia* Wilson is the oldest available subjective synonym, *Wilkingia regularis* (King, *in* de Verneuil, 1845).

MATERIAL. BM(NH): Hind Collection: Lower Limestone series, top Viséan (P2, D2 or D3, Brigantian); and Redesdale Limestone, L.-M. Viséan (B2 = D1, Asbian); Trechmann Collection: Main Limestone, Lowest Namurian, Stanhope, near Durham; and Viséan, Four Laws Limestone, Redesdale, Northumberland. EMP: de Verneuil Collection: mid-Viséan (said to be pre-Asbian), Sloboda, Toula, S. of Moscow. BGS: Redesdale Limestone, Northumberland (specimens figured by Wilson, 1959). SM: Lowick, Northumberland, Viséan.

DESCRIPTION. The shell is of medium size and very thin, resembling species of *Pleuromya* and early species of

Panopeidae except that the dorsal margins extend in a straight line further towards the posterior. It is elongate with the umbones well towards the anterior. The ventral margin is sinuous, marking an obvious but gentle subumbonal sulcus. The maximum height is at about the mid-point, well to the posterior of the umbones. The umbones are slightly tumid and only slightly raised above the hinge line. The nymphs are relatively long and very slender, barely protruding from the long, straight, adpressed posterior dorsal margins, which themselves indicate that they were joined by periostracal ligament well towards the posterior margins. The hinge is edentulous. The musculature, including the pallial sinus, is illustrated in Fig. 22.

The comarginal rugae or ribs are a little irregular, but rather constant in number from front to back. They are in the form of shell corrugations, and apart from the barely discernible growth lines are equally well preserved on the steinkern and the shell surface. There are almost imperceptible radiating ridges on the steinkerns possibly reflecting lines of surface pustules that occur across the complete shell surface except within the escutcheon. The shell has a small lunule which has no carina, instead curving imperceptibly into the anterior part of the flank. The long narrow escutcheon is limited by a sharp carina, and there is a slight shell thickening, in the form of a

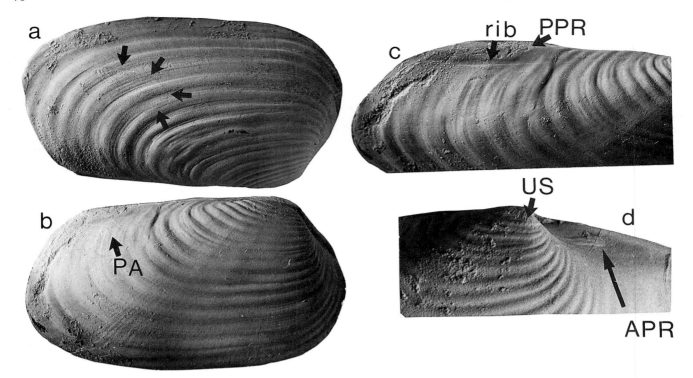

Fig. 22 *Wilkingia regularis* (King, *in* de Verneuil). Lower Carboniferous, Viséan (Asbian), Redesdale Ironstone, Redesdale, Northumberland. BM PL5002; Fig. 22a, left side, position of pallial sinus arrowed, periostracal pustules visible in postero-dorsal area, ×2; Fig. 22b, right side with posterior adductor (PA), ×2; Fig. 22c, oblique view of siphonal area of right side, showing posterior pedal retractor scar (PPR) and internal dorsal rib, approx. ×3; Fig. 22d, umbonal area viewed obliquely from top, showing scars of anterior pedal retractor (APR) and accessory umbonal scars (US), approx. ×4.

low internal rib, running at a very low angle from the umbo towards the posterior margin.

REMARKS. Hind included *Posidonomya transversa* Portlock (1843: 174; pl. 38, fig. 9) in the synonymy of this species without comment. There is no indication from Portlock's figure that it belongs to this superfamily and we consider that Hind made a mistake. Another early nominal species that belongs in *Wilkingia*, *Lutraria primaeva* Portlock (1843: 441; pl. 34. fig. 5), was curiously interpreted by Hind (1900: 307) as an *Edmondia*. Hind went so far as to claim that another specimen had been substituted for the original, but it is clear that this is not so, from both Portlock's drawing and from his measurements. However, the measurements and the figure suggest that this species falls outside the variation of *W. regularis*, and it is probably a senior synonym of *Allorisma monensis* Hind. *Wilkingia regularis* differs from *W. primaeva* in being more elongate and having a differently shaped pallial sinus.

Hind (1900: 424; pl. 48, figs 1–2) described specimens of this species from Lowick and Calderwood as *Allorisma variabilis* (M'Coy). The lectotype of that nominal species (Fig. 17, p. 69) is, however, a crushed specimen of *Cosmomya*, also from Lowick. Amongst Hind's specimens, the ribbing on the umbonal area is only significantly irregular in the lecto-type, and this is characteristic of *Cosmomya*. Most of the specimens of the present species, recognized by Hind as *Allorisma sulcata*, by Wilson as *Wilkingia elliptica* and by Astafieva-Urbaitis as *Allorismiella sulcata*, come from the Redesdale Ironstone, of high Viséan, Asbian, age, in

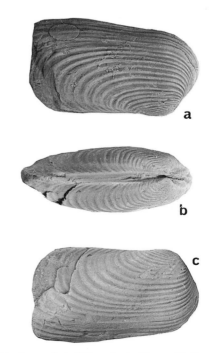

Fig. 23 *Wilkingia regularis* (King, *in* de Verneuil). Lower Carboniferous, Viséan (Asbian), Redesdale Ironstone, Redesdale, Northumberland. BM L45252, Hind Collection, **neotype** (selected herein) of *Venus elliptica* Phillips, 1836 (*non* Lamarck, 1818). Fig. 23a, left valve; Fig. 23b, dorsal view; Fig. 23c, right valve; all ×1.

Northumberland. The material available from this locality is often a little crushed, many of the specimens have eroded anterior and posterior margins, and many of the specimens are of small size. The limestone specimens from Lowick and Stanhope are better preserved and larger, as are the specimens from Sloboda in what used to be the 'Gouvernement de Toula', south of the Moscow Basin. The difference in preservation and size led Astafieva-Urbaitis to identify smaller gaping specimens as *A. sulcata*, distinct from *A. regularis* (1962: 36). We are uncertain whether or not these smaller Russian specimens belong to the same species. However, simple measurements of shell length and height (Fig. 21) suggest that the British specimens do not differ in these parameters, and they are also very close to the type material from the type locality of *Wilkingia regularis*. When Astafieva-Urbaitis introduced the name *Allorismiella* she was unaware that Wilson had at the same time been working on similar species and had introduced a new name three years earlier; she also had no opportunity to examine either Hind's material or the type series of *W. regularis*, which are the types of *Allorismiella* and *Dulunomya* respectively. She was working only with undescribed Russian specimens. Her holotype of *Dulunomya serpukhovensis* (Astafieva-Urbaitis & Dickins 1984: pl. 2, figs 1a–d) is identical in shape and sculpture to two of the paralectotypes of *Wilkingia regularis* (Figs 20e, 20f; those numbered 1744). Other specimens she refers to (Astafieva-Urbaitis, 1962: pl. 39, fig. 1) are apparently more closely related to the species described here as *Praeundulomya maxima* Portlock.

Fig. 24 *Wilkingia regularis* (King, *in* de Verneuil). Upper Carboniferous, Lower Namurian, Main Limestone, Stanhope, Weardale, Northumberland. BM PL5003; Fig. 24a, left side; Fig. 24b, right side; both ×0·9.

Genus *DYASMYA* nov.

TYPE SPECIES. *Allorisma elegans* King, 1850.

DESCRIPTION. Small to medium-sized undulomyine with sharp escutcheon and lunule. Escutcheon relatively wider than in *Wilkingia*. Subumbonal sulcus negligible or absent. Umbones more prominent than in other members of the subfamily. Anterior of shell prominent. The flatish posterior-dorsal or siphonal area is set off by a marked change in surface, forming a low rounded angle from the umbones to the posterior ventral margin, subcarinate. The shell structure of the type species is unknown but the outer surface bears very dense, close rows of small periostracal spicules. *Dyasmya* has a small but sharp pallial sinus, not much differing from that of a young *Wilkingia*. A small posterior gape is present. The comarginal rugae are less regular than those of *Wilkingia*, and in some specimens of the type species are barely present. The ligament is external, opisthodetic, parivincular, of medium length, mounted on narrow nymphs. The hinge plates are slender and parallel when viewed from above and without any teeth.

REMARKS. Beside the type species, the genus may include *Allorisma baldryi* Thomas, 1928, apparently from the Lower Permian of Peru; *Thracia longa* and possibly the more rounded *Thracia alta*, both of Lutkevich & Lobanova, 1960, from the Lower Permian of the Taimyr peninsular; and *Sanguinolites lunulatus* (Keyserling) as interpreted by those authors. Unfortunately the hinges of none of these species are known.

The more quadrate species of *Dyasmya* are very similar in outline to the Mesozoic genus *Pleuromya*, while the more rounded *Dyasmya alta* (particularly those individuals figured by Lutkevich & Lobanova (1960: pl. 11, figs 3–7)) resembles the Jurassic genus *Gresslya*. Both the Mesozoic genera have more advanced hinge types and a homogeneous inner ostracum, but they could have descended from *Dyasmya*.

The Jurassic species '*Pleuromya*' *angusta* Agassiz, 1843, commonly attributed to *Arcomya*, is very similar in shape but has many fewer rows of periostracal spicules. At present this seems to be the most suitable genus for Agassiz' species. We have observed a nacreoprismatic aragonite shell in specimens of this species in the BM(NH) from the Lias of southern England. *Arcomya* has an Upper Jurassic type species and is relatively longer and narrower, with a wide subumbonal sulcus sloping down and back below the umbones; it does not have the prominent umbones of *Dyasmya*.

Dyasmya elegans (King, 1850)   Figs 25a–b

1850 *Allorisma elegans* King: 198; pl. 16, figs 3–5.
1967 *Wilkingia elegans* (King) Logan: 63–4; pl. 10, figs 6–10.

REMARKS. The shell has very dense, closely-packed lines of small, periostracal spicules, apparently over the total exterior surface of the shell. There is clear but fragmentary confirmation of the position of the pallial line as illustrated by Logan (1967: pl. 10, fig. 6a–b) in two of our rather poorly preserved internal moulds (BM PL96 and BM PL5006). These are from the lower part of the shell limestone of the Magnesian Limestone of Claxheugh Quarry, County Durham, England, and are apparently of Kazanian age.

We have no further information concerning this species beyond that available to Logan (1967) when he was preparing his monograph. We note, however, that it may well be intermediate between Carboniferous *Wilkingia* and the Mesozoic genus *Pleuromya*. *Pleuromya* has a form of pallial line and accessory muscle scars suggesting that it had a similar form of mantle fusion and siphon formation to the living

of the dorsal shell margins typical of *Pleuromya* and other Myacea.

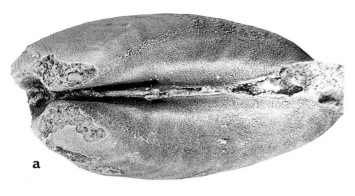

Fig. 25 *Dyasmya elegans* (King). Upper Permian, Magnesian Limestone, Claxheugh Quarry, Durham; C. T. Trechmann collection. Fig. 25a, BM PL5004, top view; Fig. 25b, BM PL5005, left side; both ×2.

Fig. 26 *Wilkingia* sp. Lower Carboniferous, Kansas. USNM, no register number; Fig. 26a, left side; Fig. 26b, ventral view; both approx. ×0·8.

genera *Mya* and *Panopea*, differing somewhat in these features from the type species of *Wilkingia*. We have been unable to find additional material with these important details preserved. It may be important that the form of the escutcheon resembles that of *W. regularis* without the lateral separation

Genus *PRAEUNDULOMYA* Dickins, 1957

TYPE SPECIES. *Praeundulomya concentrica* Dickins 1957, by original designation.

DESCRIPTION. Transversely elongate, with well-developed escutcheon behind umbones in the form of a flat marginal area. The ligament proper was apparently lodged in a relatively short groove on the proximal edge of narrow nymphs close behind the umbones. The flat marginal area was probably covered and joined by periostracum, continuous with the ligament (see p. 54). The muscle scars are very shallowly impressed and a deep pallial sinus is visible in *P. maxima* and *P. subcuneata* (but not the type species). There are one or two ribs running below the hinge posterior to the umbones. These appear as grooves on the steinkern or composite mould, and are better developed than in *Wilkingia*. In the Permian species *P. concentrica*, the two posterior grooves are as well developed as they are in *Undulomya*. In outline from above, bivalved specimens have a distinctive tapering shape.

Praeundulomya is distinguished from *Wilkingia* essentially by the shape; in *Praeundulomya* the ventral and dorsal margins are sub-parallel with only a very slight subumbonal sulcus, whereas *Wilkingia* has a much more sinuous ventral margin. This difference is less obvious in young specimens.

Fig. 27 *Wilkingia granosa* (Shumard). Upper Carboniferous, Pennsylvanian, 'Cisco, Graham', Young County, Texas. USNM Acc. 27130 (1 of 9), left side; the right valve is raised and shows the anterior part of the ligament nymph (the posterior part of the right valve is considerably eroded); ×1·3.

REMARKS. Species here included as *Praeundulomya* have mostly been attributed to *Wilkingia* or *Dulunomya* by other recent authors. In the Gondwana area *Praeundulomya* seems to have been replaced by *Undulomya* and *Exochorhynchus* early in Permian time, before the begining of the Kazanian. The transition between *Praeundulomya* and *Undulomya* was demonstrated by Dickins (1957). Transitional species such as *U. insolitus* (Thomas, 1928) therefore indicate an age somewhere in the Upper Artinskian, following the correlation of Dickins (1963: 21). In the Amatope Mountains, Peru, Permian bivalves and Pennsylvanian ammonoids seem to be associated; this might be due to mixing during collection.

Fig. 28 *Praeundulomya maxima* (Portlock). Lower Carboniferous, Viséan, Donaghenry, Co. Tyrone, Ireland; BGS 6561, holotype; Fig. 28a, top view; Fig. 28b, left valve; Fig. 28c, ventral view; all ×0·88.

Praeundulomya maxima (Portlock, 1843) Figs 28–29

1843	*Sanguinolaria maxima* Portlock: 434; pl. 36, figs 1a, 1b.
1851a	*Sanguinolites clava* M'Coy: 172.
1852	*Allorisma terminalis* Hall: 413; pl. 2, fig. 4a–b.
?1859	*Allorisma subcuneata* Meek & Hayden: 37; pl. 1, figs 10a–b.
1898	*Allorisma subcuneata* Meek & Hayden; Weller: 79–80.
1900	*Allorisma maxima* (Portlock); Hind: 419; pl. 47, fig. 5.
?1900	*Allorisma maxima* (Portlock); Hind: 419; pl. 47, figs 6, 7a, 7b.
1962	*Allorismiella regularis* (King); Astafieva-Urbaitis: 39, fig. 1.
?1974	*Wilkingia terminale* (Hall); Runnegar: pl. 1, fig. 31.
?1984	*Dulunomya maxima* (Portlock); Astafieva-Urbaitis & Dickins: 39.

HOLOTYPE. BGS (Leeds) 6561 (figured Portlock (1843: 434; pl. 36, figs 1a, 1b) and Hind (1900: pl. 47, fig. 5)), preserved in a light grey Carboniferous Limestone (possibly early Viséan), from Tyrone, Donaghenry, Co. Tyrone, Ireland.

OTHER MATERIAL. SM E1089, the type of *Sanguinolites clava* M'Coy 1851, and SM E1090, both from the Upper Grey Limestone, Upper Viséan, D2 (lower part of the Brigantian), at Llangollen, north Wales. SM E1090 is apparently slightly younger than the holotype.

DIAGNOSIS. Robust subquadrate species, with thin shell, ventral margin more or less parallel to the dorsal. Ornament of coarse, evenly rounded rugae parallel to the exterior, anterior and most of the ventral margins, but thicker and non-parallel at the posterior margin.

DESCRIPTION. Lunule and escutcheon present. Escutcheon made up of an elongate flattened area. Below the escutcheon there is a well-marked groove. The adductor muscles and the rear part of the pallial line are clearly visible in BM L47524. The anterior adductor muscle is rounded and slightly ovoid in a dorsoventral direction. Above is a small rounded scar, apparently of the anterior pedal retractor. The posterior adductor is high, rounded, but lightly marked. A deep pallial sinus is visible below the muscle scar. Thin shell is preserved in a few places. A small posterior gape was apparently present.

COMPARISONS. The specimens figured by Hind (1900: pl. 47, figs 6, 7) from the Viséan of Llangollen, north Wales, can be seen to have a slightly more sinuous posterior shell outline and have slightly less regular ribs than the holotype, even though the latter is not a complete specimen. At present we do not regard this as a specific difference. In fact specimens from the Upper Pennsylvanian of Texas named *Allorisma subcuneata* by Meek & Hayden (1858) more closely resemble the holotype. These younger specimens are usually a little smaller than *P. maxima*, but it is difficult to pick out any specific difference even in the best-preserved individuals. Also *Allorisma terminalis* Hall (1852) is probably a synonym of *P. subcuneata*. At present we do not have enough material on which to make measurements that might confirm our view that these species are similar in shape, and we tentatively include both as synonyms of *P. maxima*. Likewise we are unable to distinguish the two specimens from the Viséan of the Moscow Basin listed in the synonymy.

Praeundulomya maxima is very similar to the Permian type species *P. concentrica* Dickins 1957, but the latter has broader radiating internal posterior shell ribs. It also has comarginal low ribs that are more broadly spaced, which undergo low angular changes of direction in the lower part of the posterior or siphonal area. The pallial sinus of *P. concentrica* has not been observed although it is assumed to have been present, as an advanced character shared with *Wilkingia* and other species of *Praeundulomya*.

Genus *EXOCHORHYNCHUS* Meek & Hayden, 1865

TYPE SPECIES. ?*Allorisma altirostrata* Meek & Hayden, 1858, by original designation.

REMARKS. Examination of the type material of *E. altirostrata* in the United States National Museum led us to reject the use of this generic name, because all of the specimens were crushed and apparently foreshortened along their long axis, making it impossible to compare them with *Praeundulomya*.

Fig. 29 *Praeundulomya maxima* (Portlock). Upper Carboniferous, Upper Pennsylvanian, Cisco Formation, near Jacksboro, Texas. BM PL5012. C. H. C. Brunton Collection (previously identified as *subcuneatus* Hall); Fig. 29a, dorsal view; Fig. 29b, side view; Fig. 29c, anterior view; all ×1·5.

Runnegar (1974) also took this view and rejected the name as a *nomen dubium*. However, our more recent examination of the type material of *Allorisma barringtoni* Thomas, 1928, first described as Carboniferous but here reinterpreted as Upper Artinskian, has shown us that there really are some species that have shells much shorter than *Undulomya maxima*, and which seem to be intermediate in form between Undulomyinae and the Mesozoic species of *Homomya* and *Pholadomya*. They also show no trace of the internal dorsal posterior ribs that are typical of *Undulomya* and *Praeundulomya*. For these reasons we resurrect the generic name *Exochorhynchus*.

Exochorhynchus barringtoni (Thomas, 1928)
Figs 30a–b

1928   *Allorisma barringtoni* Thomas: 221–2; pl. 7, figs 5, 6.
?1960   *Allorisma similis* Lutkevich & Lobanova: 83; pl. 10, figs 6–8.

MATERIAL. The holotype is SM A4948 (Fig. 30), and SM A4971 is one of several paratypes in the same collection; all are apparently from the Permian, probably the Upper Artinskian, of Sullana Rd., 1·5 miles south of El Muerto and Steel Hill, Parinas Quebrada, NW Peru, but (?wrongly) associated with mid-Pennsylvanian ammonoids when they were collected.

DESCRIPTION. A medium-sized species with rounded posterior

and anterior margins and the ventral margin sub-parallel to the hinge line. The regular, low comarginal rugae are very similar to those of *Praeundulomya*. The posterior gape is very narrow and the narrow escutcheon is defined by a low, but distinct, carina. The umbones are well to the anterior, and the shell is apparently very thin.

REMARKS. *Exocorhynchus barringtoni* resembles the concentrically ribbed Pholadomyidae in the Triassic and it may be ancestral to them. There is also a considerable similarity to the Australian genus *Vacunella*. For the present, however, we follow the view of Runnegar and others that *Vacunella* developed independently in the Australasian area from some species of *Myonia* lacking a pallial sinus.

A number of similar species of *Exochorhynchus* have been described from Mongolia by Astafieva-Urbaitis (1981). The species described by Lutkevich & Lobanova (1960) from the Taimyr Peninsula of arctic Russia is somewhat distorted but has no features to distinguish it from the present species.

Subfamily CHAENOMYINAE Waterhouse, 1966
Genus *CHAENOMYA* Meek, 1865

TYPE SPECIES. ?*Allorisma leavenworthensis* Meek & Hayden, 1859, by monotypy.

Fig. 30 *Exochorhynchus barringtoni* (Thomas). ?Lower Permian, ?Upper Artinskian, Sullana Road, 1·5 miles south of El Muerta Steel Hill, Parinas Quebrada, Amotape Mountains, NW Peru. SM A4971, paratype; Fig. 30a, dorsal view; Fig. 30b, left valve; both ×1·5.

DESCRIPTION. Medium-sized, elongate shells with the umbones about half-way between the mid-point and the anterior margins. Prominent rounded posterior gape present, of almost the full shell height. Hinge without teeth; a well-developed ligament nymph extends for a short distance behind the umbones which supports a stout short C-spring ligament (see Fig. 31). The rounded edges of the escutcheon fade half-way to the posterior margins, which diverge in a gentle curve to form the top of the siphonal gape. The shell is thin and covered with regular rows of periostracal spicules.

COMPARISONS. *Chaenomya* Meek, 1865, has a broad posterior gape and hence, by comparison with living taxa, probably had long, conjoined, periostacum-covered siphons. It does not, however, have a deep pallial sinus, a feature we take to be a synapomorphy of the Undulomyinae, and we therefore consider that it lies on a separate line of descent from *Pholadomya* and that subfamily. *Chaenomya* is very similar to the Jurassic genus *Osteomya* but we think that this is a case of convergence; *Osteomya* shares the transcurrent rugae on the anterior flank with the partly contemporaneous genus *Plectomya*, which differs only in having a narrow posterior gape. *Chaenomya* has more prosogyral umbones than either *Osteomya* or *Plectomya*. There are comparable dense pustulose striae on the flank but these are much more prominent on the siphonal area of *Chaenomya* than either of the two Jurassic genera. The convergence probably reflects

comparable increase in the development of the siphons. *Chaenomya* also shows convergence, in characters we associate with deep burrowing, with the Undulomyinae and the East Australian Permian genus *Vacunella*. Both of these taxa have a prominent inflexed pallial sinus and only a very modest posterior gape, which leads us to believe they belong to a different line of descent. It is possible that *Chaenomya* evolved from a species of similar shape and with a similar pallial line but without the wide posterior gape, such as the species described by de Koninck (1885) as *Chaenomya jacunda* (see p. 82 below). The wide posterior gape seems to be an alternative strategy of siphon formation to that of *Wilkingia* and *Pholadomya*, where in the living genus at least type 'C' siphons are developed with only a modest posterior gape. *Australomya* Runnegar (1969) is more compressed, lacks the distinct posterior or siphonal area and has a tendency towards opisthocline umbones. In this last character it resembles later genera such as *Thracia* and *Plectomya*. At present we are unable to ascribe more than this one species to the subfamily. Runnegar (1974: 928–9) also included *Cosmomya* in the Chaenomyinae, because he rejected the use of the name Sanguinolitidae, following his inclusion of *Sanguinolites* in the Grammysiidae.

Chaenomya leavenworthensis (Meek & Hayden, 1859)
Fig. 31

1859   ?*Allorisma leavenworthensis* Meek & Hayden: 263–4.
1865   *Chaenomya leavenworthensis* (Meek & Hayden);
       Meek: 42.
1967   *Chaenomya leavenworthensis* (Meek & Hayden);
       Runnegar: 63; pl. 11, figs 12–13.
1969*b*   *Chaenomya leavenworthensis* (Meek & Hayden);
       Waterhouse: 38–9, figs 7J, 8I, 13; pl. 1, fig. 4; pl. 2,
       figs 5–9; pl. 3, figs 1–4, 7.
1974   *Chaenomya leavenworthensis* (Meek & Hayden);
       Runnegar: 929, text-fig. 5g; pl. 3, figs 5, 7.

MATERIAL. One specimen, USNM collection, from the Upper Carboniferous, at loc. 515g, Lower Graham Formation, 0·5 miles north of Texas 24, 6·5 miles west of Jacksboro, Texas.

REMARKS. This beautifully preserved specimen shows the distribution of spicules and the form of the ligament in perfect detail. The internal characters were well illustrated by Runnegar (1974: text-fig. 5g).

Fig. 31 *Chaenomya leavenworthensis* (Meek & Hayden). Upper Carboniferous, Lower Graham Formation, locality 515g, 0·5 miles north of Texas 24, 6·5 miles west of Jacksboro, Texas. USNM; with ligament and periostracal spicules preserved; dorsal view, approx. ×1.

Subfamily ALULINAE Maillieux, 1937

REMARKS. These are elongate shells with an extended posterior. The shell surface bears rows of prominent periostracal pustules, which we interpret as a synapomorphy of the majority of the Anomalodesmata. The ligament is external, borne on a well-defined nymph and, most importantly, a well-formed cardinal tooth is present. The hinge of *Alula* is well illustrated by Runnegar & Newell (1971).

Genus *TELLINOMORPHA* de Koninck, 1885

TYPE SPECIES. *Tellinomorpha cuneiformis* de Koninck, 1885, by monotypy.

COMMENTS. The elongate form and rudimentary cardinal tooth in the right valve link *Tellinomorpha* with the Alulinae.

Tellinomorpha cuneiformis de Koninck, 1885   Figs 32a–b

1885   *Tellinomorpha cuneiformis* de Koninck: 90–1, pl. 21, figs 1, 2.
1900   *Tellinomorpha cuneiformis* de Koninck; Hind: 433, pl. 49, figs 5–9.

HOLOTYPE. Musée nationale d'Histoire naturelle de Belgie, Brussels, no. 1698, from the Lower Carboniferous, Viséan, at Argenteau, near Visé, Belgium; this is the only known specimen.

Fig. 32 *Tellinomorpha cuneiformis* de Koninck. Lower Carboniferous, Viséan, Argenteau, near Visé, Belgium. MNHN 1698, holotype; Fig. 32a, exterior of right valve; Fig. 32b, hinge area of interior of same valve, ×1.

DESCRIPTION. The holotype has the characteristic shape of a sanguinolitid, with a broad subumbonal sulcus and a sinuous ventral margin. The posterior part of the shell is attenuated, with the narrow siphonal area demarcated by the upturn of the growth lines. The siphonal margin has a slight median sulcus shaped so that two siphonal orifices are formed. The central surface of the flank has fine radiating striae which are, however, eroded and it is not certain whether or not they bore surface pustules. Irregular surface rugae are present indicating wrinkles in the periostracum. The hinge of the right valve (the only one known) has teeth below the umbo resembling those of some heterodonts and other groups where weak 'cardinal' teeth are present. The formula is RV

(1) 0 1 0 N, which resembles some found in the trigoniacean family Schizodidae, but the arched gap or hiatus typical of the schizodids (Newell & Boyd, 1975: fig. 2) is not present, and *Tellinomorpha* has a very typical anomalodesmatid shape. The hinge is similar in the disposition of the teeth to that of *Alula* (figured by Runnegar & Newell, 1971: fig. 270) although they are less prominent, more like those of the Permophoridae. Simple teeth of this nature have apparently developed independently in a number of closely and distantly related stocks. *Tellinomorpha* does not have the elongate escutcheon typical of most Sanguinolitidae. The dorsal margin is apparently not parallel to the plane of commissure, indicating that there may have been both anterior and posterior shell gapes. There is a short, moderately stout ligament nymph behind the umbo, separated from the dorsal shell surface by a well-formed, narrow ligament groove.

REMARKS. *Tellinomorpha* may be compared with '*Sanguinolites*' *clavatus* Etheridge 1877 (*non Allorisma clavata* McChesney 1860), but that species has an elongate, carina-bound escutcheon, indicating that the periostracal ligament joined the two valves back to the dorsal posterior margin, whereas the dorsal margins of *Tellinomorpha* apparently diverged posteriorly in a similar fashion to *Chaenomya* (Fig. 31). Examination of specimens from the Viséan Limestones of the Derbyshire Dome, England, attributed to *Tellinomorpha* by Hind (1900), show that these do not have the attenuated siphonal area of this genus and belong, in fact, to *Wilkingia*.

Subfamily Uncertain

An unnamed genus, intermediate in form between *Gilbertsonia* of the Sanguinolitinae and *Chaenomya*, is described here. It is possibly ancestral to *Chaenomya* or *Vacunella* or to both, and it includes the species '*Chaenomya*' *jacunda* de Koninck. Without further details of its characters we are uncertain in which subfamily it should be included.

Genus Uncertain

'*Chaenomya*' *jucunda* de Koninck resembles *Gilbertsonia* in shape but has a broad, shallow pallial sinus, somewhat resembling that of several species of *Myofossa* and *Cosmomya*. A narrow but obvious siphonal gape is present, which is about two-thirds of the total shell height, and therefore not as extensive as the posterior gape of *Chaenomya leavenworthensis*.

'*Chaenomya*' *jacunda* de Koninck, 1885   Figs 33a–d

1885   *Chaenomya jacunda* de Koninck: 7; pl. 1, figs 1–8.
1974   *Sedgwickia? jacunda* (de Koninck) Runnegar: pl. 3, figs 14–16.

MATERIAL. BM L13446, BM L13481, BM L47500 and BM PL1655, from the Lower Carboniferous, Viséan, at Tournai, Belgium.

DESCRIPTION. Medium-sized anomalodesmatid, gibbous with forward-pointing umbones well to the anterior. A small lunule is present with a subrounded carinate margin. The posterior dorsal margin is relatively long and straight, set in a broad escutcheon which is bounded by subrounded carinae. The siphonal margins are relatively long and straight, forming

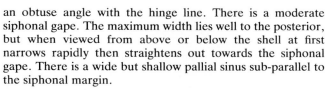

Fig. 33 *'Chaenomya' jacunda* de Koninck. Lower Carboniferous, Tournasian, Tournai, Belgium. Figs 33a–c, BM PL1655; Fig. 33a, ventral view; Fig. 33b, top view; Fig. 33c, right side (note the shape of the pallial line); all ×0·8. Fig. 33d, Upper Carboniferous, ?Texas; USNM 5953; ×1·4.

an obtuse angle with the hinge line. There is a moderate siphonal gape. The maximum width lies well to the posterior, but when viewed from above or below the shell at first narrows rapidly then straightens out towards the siphonal gape. There is a wide but shallow pallial sinus sub-parallel to the siphonal margin.

REMARKS. '*C.*' *jacunda* resembles some species of *Myonia* and *Vacunella* in shape and it is one possibility that it is their ancestor. An example (USNM 5952, Fig. 33c), possibly different, from the Upper Pennsylvanian of Texas, suggests that this rather rare taxon may have had a considerable time range. '*C.*' *jacunda* is also closely similar to some species of the Mesozoic genus *Pachymya*, particularly *P. crassiuscula* (Morris & Lycett, 1855) from the Bacocian of Normandy and England. '*C.*' *jacunda* is similar to *Myofossa omaliana* (de Koninck) in general shape and in the form of the pallial line. It is, however, much larger and does not have the opisthodetic umbones typical of *Myofossa*. The two could share close common ancestry.

Family **PERMOPHORIDAE** van de Poel, 1959
[*Pro* Pleurophoridae Dall, 1895; I.C.Z.N., Art. 40]

In the present paper we propose that the family Permophoridae should be included with the Anomalodesmata rather than with the heterodonts, as in the *Treatise* classification (Chavan, *in* Moore, 1969: N543). Aspects of the *Treatise* diagnosis of this family are, we suggest, an interpretation, based on its assumed systematic position. The following phrases, quoted from the *Treatise*, are clearly correct if the Permophoridae were properly interpreted as carditaceans; in fact they are an expression of the more obvious differences: 'Cardinals partly obsolete', 'radial ribs tending to be obsolete on anterior part of surface', 'anterior laterals lacking in most' (Chavan,

in Moore, 1969; N543). The similarities between the Permophoridae and the Carditidae and the placing of the Permophoridae in the Carditacea depend upon the interpretation of the dentition. In heterodont terms, the teeth of the Permophoridae may be described as lucinoid when they are present. Bernard's analysis of heterodont teeth (1895), although of immense value in the Veneroida, is now suspected of supporting false homologies in the lucinoids (MacAlester 1966, Morris 1978). The considerable doubt concerning homologies of teeth between the heterodonts and the Trigoniacea had led Boyd & Newell (1968) to abandon the Bernard-Douvillé system for that superfamily and instead make use of a more objective (i.e. with no presupposition of homology) Steinmann notation. We suggest that the Permophoridae and the Carditidae belong to quite different subclasses of bivalves which separated before or at the very beginning of the Ordovician. If by some chance their cardinal teeth are homologous, which is unlikely on our present evidence, their form would be primitive for the two subclasses and would not indicate a close relationship between the two families. We think that the apparent similarity of tooth pattern in the two families is more likely to be the result of convergence. The number of possible teeth patterns is limited when there are fewer individual teeth. We therefore urge the use of the Steinmann notation for toothed forms of the Permophoridae. The true relationship of the Carditacea, we consider, has been recognized by Yonge (1969), who described the great similarity of 'mantle fingers' between the teeth in both Carditidae and Astartidae. The modioliform shape of some Maastrichtian to Recent Carditidae we would interpret as an advanced character because nearly all the earlier Cretaceous Carditidae are round in shape, resembling *Cyclocardia* and *Venericardia* and similar or more gibbous species of Astartidae, but with radiating ribs.

There is no clear evidence in the fossil record that the Cretaceous to Recent Carditidae are descended from those in the late Triassic. The fact that the Cretaceous carditids share most of their characters with the Astartidae leads us to believe that the Carditidae as presently recognized (Chavan 1969) are not a single clade. We believe that more than one group descended from ancestors at present classified with the Astartidae. They are, we believe, essentially 'Astartidae' which have developed radial ribs, a phenomenon that probably happened more than once. We therefore support Yonge's view that there is no need for two superfamilial names, Crassatellacea and Carditacea.

Most Permophoridae may be distinguished from most Mytiloida by the presence of a clearly-marked escutcheon and the fact that the external ligament is mounted on relatively short, upward facing, nymphs although there is considerable overlap between the two taxa in overall shell shape.

Subfamily **PERMOPHORINAE** van de Poel, 1959 (1895)
[*Nom. trans.* Chavan, *in* Moore 1969]

Genus *PERMOPHORUS* Chavan, 1954
Figs 34–35

TYPE SPECIES. *Arca costata* Brown, 1841, by monotypy.

REMARKS. The nomenclature of this genus is fully dealt with by Chavan in the *Treatise* (Chavan, *in* Moore 1969: N543). Examples of the type species and a similar species from the Permian of the Glass Mountains are figured here to show clearly the characteristics of the genus.

Fig. 34 *Permophorus costatus* (Brown). Upper Permian, Magnesian Limestone, England; BM PL235, steinkern; Fig. 34a, right side; Fig. 34b, left side; both ×1·5.

Licharew (1925: 125) proposed a new name, *Pleurophorina*, in which he included a single species, *Modiola simpla* Keyserling (1846: 28; pl. 10, fig. 22; pl. 14, fig. 1), which must, therefore, be the type of the genus by monotypy. The hinges figured by Licharew (1925: pl. 1, figs 1, 2) are close to *Permophorus* Chavan (1954) (pro *Pleurophorus* King, 1844, *non* Mulsant, 1842). If Licharew's specimens are correctly referred to *Modiola simpla* Keyserling, and we are not able to check on this, then *Permophorus* may be a synonym of *Pleurophorina*.

The shell is elongate-ovate with the umbones well towards the anterior. The rounded posterior margin and the area from

it to the umbones is separated from the flank by a fine radial rib. There are sometimes further radial ribs on this posterior dorsal area. The flank and anterior are usually smooth. The hinge has an opisthodetic parivincular ligament set on slender short nymphs. Posterior lateral teeth occur, sometimes in each valve. The right valve has a single, moderately large cuneiform tooth which fits between the two subumbonal teeth of the left valve. These may be seen in the rather badly preserved steinkerns (Logan 1964) of *P. costata* but are better observed in *Permophorus* cf. *albequus* (Beede) (Fig. 35). The anterior adductor scar is of small to medium size, and deeply inset leaving a well-formed vertical buttress behind it. The posterior adductor is rounded and of medium size and set below the distal end of the posterior lateral tooth. The pallial line is entire although it is usually very faint towards the posterior of the shell.

Genus *PLEUROPHORELLA* Girty, 1904

TYPE SPECIES. *Pleurophorella papillosa* Girty, 1904, by original designation, from the Graham formation, Pennsylvanian (Cisco), of Young County, Texas.

SYNONYMS. *Eopleurophorus* Elias, 1957: 780 (type species, *Cypricardia? tricostata* Portlock (1843: 441; pl. 34. fig. 17) as interpreted by Hind (1900: 391), by original designation); from Carnteel, County Tyrone and Drumkeeran, County Fermanagh, Ireland.

DIAGNOSIS. Transversely elongate, distinct lunule and escutcheon, radiating ornament fairly well developed, especially in posterior part of shell. Granulation of shell surface by periostracal spicules distinct, apparently absent in some. More or less edentulous. Ligament lodged in a narrow elongated groove in the anterior part of a flat escutcheon which extends well towards the rear. Nymphs slender and low. The anterior adductor scar is well differentiated and bounded at the rear by a distinct buttress.

REMARKS. Chavan (1969: N546) placed *Pleurophorella* in the Permophoridae, though he expressed some doubt. We refer to this genus a number of Carboniferous species which have usually been referred to *Sanguinolites*; these include *Sanguinolites tricostatus* (Portlock, 1843), which is the type species of *Eopleurophorus* Elias (1957: 781), *S. striatolamellosus* (de Koninck, 1842), *S. striatus* Hind, 1900, *S. striatogranulatus* Hind, 1900, *S. visetensis* (de Ryckholt, 1847), *S. oblongus* Hind, 1900, *S. roxburgensis* Hind, 1900, and *S. ovalis* Hind, 1900. Other Carboniferous species are also included here.

Poor development or absence of teeth, together with the granulation of the shell surface, have perhaps hindered recognition of the relationships of this group. Genera of Permophoridae with well-developed lateral and cardinal hinge teeth share with the edentulous Carboniferous forms the distinctive lunule and escutcheon and the distinctly separated anterior adductor scar. The genus *Stutchburia* from the Lower to Upper Permian (see Dickins, 1963: 95) has poorly developed cardinal teeth and variable development of posterior lateral teeth, and occupies an intermediate position. The development of external granulation (pustules—a short rounded form of periostracal spicules) is apparently variable in both edentulous and tooth-bearing forms. Its presence or absence may also reflect preservation. In Permian and

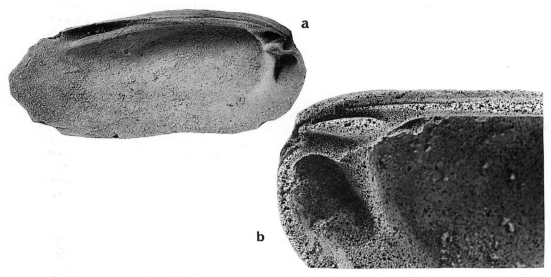

Fig. 35 *Permophorus* cf. *albequus* (Beede). Permian, West Texas; Fig. 35a, inside of left valve, ×4; Fig. 35b, umbonal area inside right valve, *c.* ×8. Photographs kindly provided by Professor N. Newell.

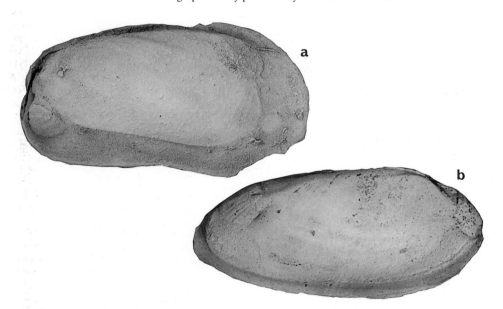

Fig. 36 Two species of *Stutchburia* with normal and unusual pallial muscle attachment. Fig. 36a, *Stutchburia farleyensis* (Etheridge); Lower Permian, Farley, New South Wales, Australia; BM PL603, with deeply inserted adductors and entire pallial line. Fig. 36b, '*Stutchburia*' sp.; Lower Permian, Bowen Coalfield, Queensland, Australia; BM PL539, with similar adductors and pallial line, but with an area of small spots, apparently of muscle attachment, below the posterior adductor, where a pallial sinus is found in many other bivalves; both ×1·5.

Triassic forms, granulation has been rarely recorded. Newell (1940: 298; pl. 3, fig. 16), however, described and figured irregularly occurring pustules in *Permophorus albequus* (see Fig. 35) from the Upper Permian of the USA, and Licharew (1925: 125) described granulation in *Pleurophorina* from the Kazanian of the USSR. From these data, it is reasonable to conclude that the forms with heterodont-like dentition are related to edentulous forms, and at present it appears that the edentulous Carboniferous forms are the more primitive in this respect.

Hind (1904) included two British Viséan species in the genus *Spathella* Hall 1885, '*Spathella*' *tumida* Hind and '*Spathella*' *cylindracea* (M'Coy). These seem to us to belong to *Pleurophorella*. Hinge details of *Spathella* are not well known, but Pojeta, Zhang & Yang (1986: 73) diagnosed the genus based on its type species as a lithophagiform modiomorphid with coarse comarginal ornament. This, and their illustrations of *Spathella typica* Hall, suggest to us that *Spathella* is not an anomalodesmatid and the two species were incorrectly placed in it by Hind.

Pojeta, Zhang & Yang (1986: 86; pl. 57, figs 5–8) also figured topotype material of *Sphenotus arcaeformis* Hall & Whitfield, 1869, the type species of that genus. They also followed Driscoll (1965) and others in ascribing to *Sphenotus* some species that we would attribute to *Pleurophorella*. When the hinge and musculature of *Sphenotus arcaeformis* are known it may well prove correct to synony-mize *Pleurophorella* with *Sphenotus*; in the meantime we prefer to use *Pleurophorella*, where these characters are now known.

Pleurophorella papillosa Girty, 1904   Fig. 37

1904   *Pleurophorella papillosa* Girty: 729–32; pl. 45, figs 4–6; pl. 46, fig. 5.
?1969 *Pleurophorella papillosa* Girty; Chavan: N546.

MATERIAL. A single specimen, USNM G. A. C. Collection, from the Upper Pennsylvanian ('Upper Finis'), hills 0·5–1

Fig. 37 *Pleurophorella papillosa* (Girty). Upper Carboniferous, Upper Pennsylvanian, Upper Finis Shale; hills 0·5–1 mile north of a point 0·3 miles NE of intersection of old Chico Road, 3·2 miles east of Jacksboro, Texas; USNM, G.A.C. Collection; dorsal view with ligament in place; *c.* ×1.

mile north of a point 0·3 miles north-east of intersection of old Chico road and Wizard Wells Road, 3·2 miles east of Jacksboro, Texas.

DIAGNOSIS. Shell surface with fine, close-packed pustules all over, otherwise without ornament. Nymphs slender and long, nearly half the length of the escutcheon.

REMARKS. We figure a specimen (Fig. 37) that conforms to Girty's original description and comes from the same area and horizon. It shows the nature of the nymph, occupying the anterior part of the escutcheon, and the surface is covered with fine close-packed pustules which seem to us to be merely low rounded periostracal spicules that do not show any particular alignment.

Pleurophorella tricostata (Portlock, 1843)   Figs 38a–i

1843   *Cypricardia? tricostata* Portlock: 441; pl. 34, fig. 17.
1900   *Sanguinolites tricostatus* Portlock; Hind: 391–3.
?1900 *Sanguinolites striatogranulatus* Hind: 393–4; pl. 42, figs 16–22.

HOLOTYPE. BGS 14747 (Figs 38a, b).

OTHER MATERIAL. BM L13446, Carboniferous Limestone, Britain (no further details recorded); BM 22545 and BM L13481, J. Wright Collection, Carboniferous Limestone, Little Island, County Cork; BM L47500, Hind Collection, Poolvash, Isle of Man; all are from the Viséan. BM L24821–3 (Gilbertson Collection no. 97) are three syntypes of *Cypricardia glabrata* Phillips, 1836, that could be young

Fig. 38 *Pleurophorella tricostata* (Portlock). Lower Carboniferous, Viséan. Figs 38a–b, County Fermanagh, Northern Ireland; BGS 14717, holotype; Fig. 38a, left valve, side view; Fig. 38b, left valve, dorsal view; both ×1·6. Figs 38c–h, three syntypes of *Sanguinolites striatogranulatus* Hind; Figs 38c–f, Poolvash, Isle of Man; Figs 38c–d, BM L47500; Fig. 38c, right side; Fig. 38d, left side; both ×1; Figs 38e–f, BM L47502; Fig. 38e, right side; Fig. 38f, left side; both ×1; Figs 38g–f, Stebden Hill, Yorkshire, England; BM L47499, with shell preserved; Fig. 38g, dorsal view; Fig. 38h, right valve; both ×0.9. Fig. 38i, Little Island, County Cork, Ireland; BM L24545, ×1·8.

individuals of this or a number of other British species. As we are unable to decide to which of the smoother species they belong, we provisionally reject Phillips' species *Cypricardia glabrata* as a *nomen dubium*.

DESCRIPTION. Both the lunule and the escutcheon are defined by a distinct carina. The escutcheon is two-thirds of the shell length and the distance from the umbones to the posterior margins is five-sixths of the shell length. The dimensions of specimen BM L24545 are: L 42 mm, H 28·5 mm, U-P 35 mm. The umbones are prosogyral and confluent with a convex dorsum on both valves, whereas *Sanguinolites* is concave in this area. The ligament nymph is visible in BM L13446 and BM L13481; it is separated from the dorsal edge of the escutcheon by a distinct narrow groove, and its dorsal margin lies just below the margin of the escutcheon.

The internal surface is visible in BM L47500. The anterior adductor is well impressed into a raised area of the inner shell surface. The pallial line is partly visible in this specimen but is very faint; it does not have a sinus. Two radiating, fine, low ribs are present between the escutcheon and the low radiating line delimiting the corselet, which distinguishes the species from *Pleurophorella visetensis* de Rychkholt (*sensu* Hind, 1900) which has three such ribs.

Closely related species which are not thought to be synonyms are: *Sanguinolites striatolamellosus* de Koninck, *sensu* Hind (1900: 398; pl. 43, fig. 11 & 11a) (Fig. 39) (*non Cypricardia striatolamellosa* de Koninck, 1842: pl. H, fig. 8a–c); *Isocardia transversa* de Koninck (1842: pl. 1, fig. 3a–b); *Sanguinolites oblongus* Hind (1900: pl. 43, figs 6–7); and possibly *Sanguinolites visetensis* (de Rychholt, 1847) *sensu* Hind (1900: 395; pl. 43, figs 1–4). We consider, however, that *Sanguinolites striatogranulatus* Hind (1900: 393; pl. 42, figs 16–22) may well be a synonym and the differences in shape and granulation may reflect preservation rather than specific differences.

Fig. 39 *Pleurophorella striatolamellosa* (de Koninck). Lower Carboniferous, Viséan, Stebden Hill, Yorkshire; BM L47510; Fig. 39a, left valve; Fig. 39b, dorsal view; Fig. 39c, anterior view; all ×1·25.

Pleurophorella sp.   Fig. 40

MATERIAL. USNM 515g, two specimens from the Lower Graham Formation, Upper Pennsylvanian, 0·5 miles north of Texas 24, 6·5 miles west of Jacksboro, Texas.

DESCRIPTION. Shell ornament consisting of concave-upwards, sharp, comarginal ribs with intervening fine growth laminae. No surface pustules are preserved. Carinate lunule and escutcheon present. The carina bounding the escutcheon is crenulate with prominent backward-pointing growth lines. The hinge of the right valve is particularly well preserved (Fig. 40); there is a small anterior tooth, parallel to the anterior dorsal margin, and a thin posterior tooth running parallel to and below the ligament nymph. The nymph is moderately long. It is not well preserved at its proximal end, where there may have been an attachment of the outer anterior ligament. The ligament groove is well preserved, and the ligament is present at the distal end; more proximally the upper surface of the nymph is transversely striate where part of the inner ligament layer has broken away. Growth laminae are visible on the inner surface of the nymph just below the attachment area of the inner ligament.

Fig. 40 *Pleurophorella* sp. Upper Carboniferous, Upper Pennsylvanian, Lower Graham formation, 0·5 miles north of Texas 24, 6·5 miles west of Jacksboro, Texas; USNM 515g, external view of left valve together with hinge of right valve; ×1·25.

REMARKS. *Pleuphorella* sp. is closely similar in morphology to *Stutchburia*.

Pleurophorella transversa (de Koninck, 1842)   Fig. 41

1842   *Cypricardia transversa* de Koninck: 94; pl. 1, fig. 3; pl. 3, fig. 8.
1842   *Isocardia transversa* de Koninck: pl. 1, fig. 3 only.
1885   *Sanguinolites transversus* (de Koninck); de Koninck: 76; pl. 17, figs 4–5.

MATERIAL. BM 32908, de Koninck Collection, is a single specimen associated with an original label in de Koninck's

Fig. 41 *Pleurophorella transversa* (de Koninck). Lower Carboniferous, Tournaisian, Tournai, Belgium; BM 32908, de Koninck Collection, ?syntype, slightly oblique view of right side to show posterior lateral tooth of left valve in its correct orientation; ×2·5.

hand reading 'Cypricardia transversa de Kon.' It is similar in shape and proportions to de Koninck's original figue, but with the valves slightly displaced, and it might be the figured syntype. A second specimen, a steinkern, bearing the same number appears to have been misidentified.

REMARKS. The displacement of the two valves has now exposed part of the hinge of the left valve (Fig. 41), which has a long, narrow hinge plate set into a long narrow carinate escutcheon. There is a slender short 'posterior lateral tooth' that we believe to have developed independently from similar teeth in the heterodonts. There is also a short nymph exposed which runs for a short distance behind the umbones, separated from the shell surface of the escutcheon by a marked ligament groove.

?*Pleurophorella cuneata* (Phillips, 1836)   Fig. 42

1836   *Nucula cuneata* Phillips: 210; pl. 5, fig. 14.
1897   '*Nucula*' *cuneata* Phillips; Hind: 205.

HOLOTYPE. BM 97147, Viséan, Bolland, Yorkshire; Gilbertson Collection.

REMARKS. This tiny specimen has umbones at the anterior where it is cordate in section. The dorsal and ventral margins diverge slightly so the greatest height is towards the posterior. The posterior margins are rounded. The shell is very similar in form to a date mussel, except that there is a clearly marked elongate carinate escutcheon, typical of the Permophoridae. The similarity of form to a date mussel raises the possibility that ?*Pleurophorella cuneata* was also a rock borer.

Fig. 42 ?*Pleurophorella cuneata* (Phillips). Lower Carboniferous, Viséan, Bolland, Yorkshire; BM 97147, Gilbertson Collection, holotype; approx. ×8.

Genus *BOWLANDIA* nov.

TYPE SPECIES. *Cypricardia rhombea* Phillips, 1836.

ETYMOLOGY. The generic name is derived from the Forest of Bowland, Yorkshire, an area locally famous for its Carboniferous fossils.

SYNONYM. *Ivanovia* Astafieva-Urbaitis, 1978 (*non* Dubrolyubova, 1935), type species, *I. slovenica* Astafieva-Urbaitis, 1978, by monotypy. See below.

NOMEN DUBIUM. *Digonomya* Whidborne 1897: 16–17 (type species, *D. elegans*) is superficially similar to the present genus. The type material is in the BGS collections but shows none of the characters of the hinge and is not well preserved; on this account we reject the generic name *Digonomya* as a *nomen dubium*.

DIAGNOSIS. The new genus resembles *Pleurophorella* but it

has a relatively reduced anterior ventral margin giving it an overall modioliform appearance; we interpret from this that *Bowlandia* had a semi-infaunal byssate to epifaunal byssate habit, much like living *Mytilus edulis*, except that we have no evidence that the genus occurred intertidally. The anterior-ventral margins converge at about 30° in the type species but at a much greater angle, about 150°, in *B. angulata*, suggesting attachment of the latter to harder and more planar substrates. The adductor muscles are somewhat anisomyarian with the anterior one varying from rather small to somewhat reduced. We have only observed the posterior adductor in *B. angulata*; it is medium-sized and rounded, and set close below the distal end of the escutcheon.

REMARKS. *Bowlandia* gen. nov. differs from *Goniophora* Phillips, 1848, in having a thicker shell, a more substantial hinge, and a sharp carinate, elongate escutcheon. It does not have the wide flange delimiting the posterior dorsal area that is present in *Goniophora*, nor the thin diverging internal buttresses that occur behind the umbones of that genus. It is most likely that the two genera belong to different super-families. *Goniophora* has recently been under investigation by Dr John Pojeta jr and may not be correctly placed in the Modiomorphidae. There are also no characters yet observed that link *Goniophora* conclusively with the Anomalodesmata.

Hind (1899: 338) proposed the name *Mytilomorpha* as a replacement name for *Goniophora* Phillips, 1848, because *Goniophorus* had been used by Agassiz for a genus of crinoids. This was, however, unnecessary. From Hind's statement *Cypricardia cymbiformis* J. de C. Sowerby is to be regarded as the type species of both *Goniophora* and *Mytilomorpha*, and therefore *Mytilomorpha* is an objective synonym of *Goniophora*.

We consider that the Carboniferous species *Bowlandia rhombea* (Phillips) and *B. angulata* (Hind) should not be placed in *Goniophora*, and belong in fact to the Permophoridae. The similarity is superficial: where *Goniophora cymbiformis* occurs in a badly crushed condition, as in the Upper Silurian of the Ludlow area, it often occurs with a second, apparently unnamed species of similar size and shape but without the considerable flange, which is probably correctly placed in *Cosmogoniophorina* Isberg, 1934. The two have been sometimes thought to be the same species, but in our opinion, this second, Upper Silurian species and the genus *Cosmogoniophorina* itself, belong to the family Permophoridae.

Bowlandia may prove to be closely related to the Upper Permian genus *Naiadopsis* Mendes, 1952 (Runnegar & Newell, 1971: 56–7, fig. 25) from the Parana Basin. *B. rhombea* also closely resembles '*Ivanovia*' *slovenica* Astafieva-Urbaitis (not *Ivanovia* of Dubrolyubova), which is, however, only known from two views of the right side. *Bowlandia slovenica* apparently has a slight flexure of the pallial line, just below the posterior adductor.

Bowlandia rhombea (Phillips, 1836)   Figs 43a–h

1836   *Cypricardia rhombea* Phillips: 209; pl. 6, fig. 10.
1885   *Sanguinolites rhombea* (Phillips); de Koninck: pl. 15, fig. 28.
1899   *Mytilomorpha rhombea* (Phillips) Hind: 338; pl. 38, figs 6–11.

LECTOTYPE. BM L3480, here designated, is the specimen

Fig. 43 *Bowlandia rhombea* (Phillips). Figs 43a–e, Lower Carboniferous, Viséan, Bolland, Yorkshire; Fig. 43a, BM L3480, Gilbertson Collection, lectotype, side view of right valve; Figs 43b–d, BM 97182, similar to the lectotype; Fig. 43b, right side; Fig. 43c, dorsal view; Fig. 43d, posterior view; all ×1·5; Fig. 43e, PL5010, probable paralectotype, Gilbertson Collection (no. 97a), view of left side; ×2. Fig. 43f, Lower Carboniferous, Viséan, Poolvash, Isle of Man; BM L47456, large elongate specimen, left side of steinkern; ×0·67. Figs 43g–i, Lower Carboniferous, Tournaisian, Tournai, Belgium; BM PL5011; Fig. 43g, anterior view; Fig. 43h, dorsal view; Fig. 43i, left valve; *c*. ×2.

figured by Phillips, from the Carboniferous Limestone, Viséan, at Bolland, Yorkshire; Gilbertson collection.

PARALECTOTYPE. BM 97182, from the same horizon and locality as the lectotype. Three further specimens from Bolland in the Gilbertson Collection (no. 97a) are listed as *Cypricardia glabrata* Phillips and therefore are unlikely to be syntypes. (BM L24821–3, Gilbertson Catalogue no. 97, are three syntypes of the true *Cypricardinia glabrata*. They are clearly not the same as those of no. 97a; although details of the hinge are not shown, they may belong to *Pleurophorella*).

OTHER MATERIAL. We refer the following specimens to *B. rhombea*:

Hind Collection: BM L45931–4, Castleton, Derbyshire; BM L45935–7, Elbolten, Yorkshire; BM L47451, Wetton Hill, Leek, Staffordshire; BM L47452–4 and BM L47456, Poolvash, Isle of Man.

Roscoe Collection: BM L43647, Wetton Hill, Leek, Staffordshire; BM L43616–26 and BM L43648, Narrowdale, Hartington, Derbyshire.

Butler Collection: BM L8175, Wetton Hill, Leek, Staffordshire.

Bather Collection: one specimen, Viséan, D_2, Peakhill Farm, Mam Tor, Derbyshire.

OTHER POSSIBLE SYNONYMS. De Koninck (1885) figured many specimens from the Carboniferous of Belgium under many new specific names. All were attributed to *Sanguinolites*, and most are closely similar to *Bowlandia rhombea*, though at the small sizes of most of them it is difficult to distinguish *Bowlandia* from the more carinate species of *Pleurophorella*. The following de Koninck species from the Viséan, étage 3, mostly from Visé, are probably synonyms of *Bowlandia rhombea*: *Sanguinolites apertus* de Koninck (1885: pl. 15, figs 1, 2), *S. solitarius* (pl. 15, figs 16, 17), *S. vexillum* (pl. 15, figs 19, 31, 32), *S. reversus* (pl. 15, fig. 25), *S. bipartitus* (pl. 15,

Fig. 44 *Bowlandia* sp. Figs 44a–b, Upper Carboniferous, Pennsylvanian, Missouri Series, Ochelata Group, Wann Formation; old uncompleted railroad cutting, 4 miles north and 2 miles east of Copan, Oklahoma, USA; USNM 6832, Conlin Collection; Fig. 44a, dorsal view; Fig. 44b, right valve. Figs 44c–d, Upper Carboniferous, Pennsylvanian, Canyon Series, Graford Group Shale, above Willow Point Island Member; Bridgeport Clay Pit, Wise County, Texas; USNM 7277, Conlin Collection; Fig. 44c, right valve; Fig. 44d, dorsal view. All slightly enlarged.

fig. 27), *S. quadricostatus* (pl. 15, fig. 34) and *S. reniformis* (pl. 15, figs 45, 46). The first specimen of *S. apertus* (pl. 15, figs 1, 2) is identical to *Bowlandia rhombea* except for the small rounded anterior gape which may have been for a byssus, but a second specimen (de Koninck, 1885: pl. 15, figs 3, 4) is much more elongate and may belong to a separate species. Other specimens from the same horizon and locality figured by de Koninck, but with de Rykholt specific names, that are probably also examples of *Bowlandia rhombea* are: *S. fabalis* (pl. 15, fig. 35), *S. praesectus* (pl. 15, fig. 37), *S. scapha* (pl. 15, fig. 38), *S. lyellianus* (pl. 15, fig. 39) and *S. tabulatus* (pl. 15, figs 41–4). Five more de Koninck (1885) species from other horizons and localities in Belgium that are probably also synonyms of *Bowlandia rhombea* are: *S. cuneatus* (pl. 16, figs 14, 15), *S. constrictus* (pl. 16, fig. 17), *S. angulatus* (pl. 16, fig. 18), *S. deletus* (pl. 16, fig. 19), all from étage 2, and *S. parvulus* (pl. 16, figs 20–3) from the Tournaisian.

COMPARISONS. *B. rhombea* is intermediate in form between *Permophorus* and *B. angulata* (Hind); the carina separates the flank from the posterior dorsal area at an angle of about 110° in *B. rhombea* but at only about 90° in *B. angulata*. The latter species, only recorded from the Viséan of Thorpe Cloud, Derbyshire, is also very much larger in all known examples except one, and has no trace of the radial cord on the posterior dorsal area close above the carina which is usually visible in *B. rhombea*.

Genus *SILIQUIMYA* nov.

TYPE SPECIES. *Sanguinolaria plicata* Portlock, 1843.

DESCRIPTION. Elongate, narrow genus of a similar shape to the Recent Solenacea genera *Siliqua* and *Cultellus*. Umbones well towards the anterior, shell thin with a slightly backwards-sloping sulcus in young growth stages only. The ligament and nymphs start between the umbones; they are opisthodetic and

parivincular, long and straight. The dorsal margins are in juxtaposition from the umbones to the posterior margins. The nymphs are narrow and elongate with a narrow ligament groove. The ligament is set in a long, narrow escutcheon limited by sharp carinae.

The shell has a posterior inner rib at a very low angle to the hinge, which appears as a sulcus on the steinkern. The shell surface has low, rounded, comarginal rugae with no clearly defined corselet, although the rugae become irregular between the siphonal margins and the umbones. No surface pustules have been observed. The pallial line is very faint except close to the anterior adductor. The specimen illustrated in Fig. 45a has a relatively small, rounded and very faint posterior adductor scar and has been interpreted by Hind as having an entire pallial line; it is almost impossible to see the posterior part of the pallial line in a number of very well-preserved specimens, but none has a visible pallial sinus. The anterior adductor scar is rounded and slightly truncated towards the umbones. It is well inserted in front of a moderately thick buttress. There is a prominently inserted anterior pedal retractor between the anterior adductor and the umbones, lying close to the hinge. Small accessory muscle scars form a group of short incised striae on the anterior surface of the umbones of the steinkern. The shell appears to have a slight posterior gape (but see Hind, 1900: 389).

DISCUSSION. From M'Coy to the present time, *Siliquimya plicata* has always been placed in *Sanguinolites*. It differs from that genus, however, in having a more gently rounded posterior margin with no clearly defined corselet and in being less gibbous. It resembles a much elongated version of the Devonian genus *Glossites*. We have been influenced by the outline shape of ?*Pleurophorella striata* (Hind, 1900, 401–2; pl. 46, figs 1–2, & pl. 50, fig. 22), intermediate between *Siliquimya* and *Pleurophorella* of the Permophoridae. This leads us to suggest that *Siliquimya* should be included in the Permophoridae and its similarity to taxa included in the

Sanguinolitinae is partly a result of parallelism. Unfortunately the detailed characters of the accessory musculature of ?*Pleurophorella striata* that might corroborate this interpretation are as yet unknown.

Siliquimya plicata (Portlock, 1843)   Figs 45a–c

?1842   *Sanguinolaria plicata* M'Coy, *in* Griffith: 12. *nomen nudum.*

1843   *Sanguinolaria plicata* Portlock: 433; pl. 34, fig. 18.

1843   *Sanguinolaria transversa* Portlock: pl. 34, fig. 21 [see discussion of *Wilkingia*, p. 73]

1844   *Sanguinolites plicatus* (Portlock); M'Coy *in* Griffith: 49; pl. 10, figs 3a, 3b.

1844   *Sanguinolites iridinoides* M'Coy *in* Griffith: 49; pl. 12, fig. 1.

1849   *Pholadomya iridinoides* (M'Coy); d'Orbigny: 128.

1900   *Sanguinolites plicatus* (Portlock); Hind: 387–8; pl. 44, figs 9, 11–15; pl. 45, figs 1–4.

?1900   *Sanguinolites striatus* Hind: 401; pl. 50, fig. 22 only.
[Fuller synonymies were given by Hind (1900: 387–8) and Paul (1941)].

TYPE MATERIAL. The holotype in the BGS collections, figured by Hind (1900: pl. 44, fig. 11), is a young individual with both valves preserved. The holotype of *Sanguinolaria transversa* Portlock, in the same collection, was also figured by Hind (1900: pl. 45, fig. 1). SM E1045, here designated the lectotype of *Sanguinolites iridinoides* M'Coy, is the specimen figured by M'Coy, and is from the Viséan of Lowick, Northumberland.

OTHER MATERIAL. BM PL2760, from the Viséan near Keswick, Northumberland. BM L5224, BM L28181–8, BM L46457 and BM L46473–6 from the Redesdale Ironstone,

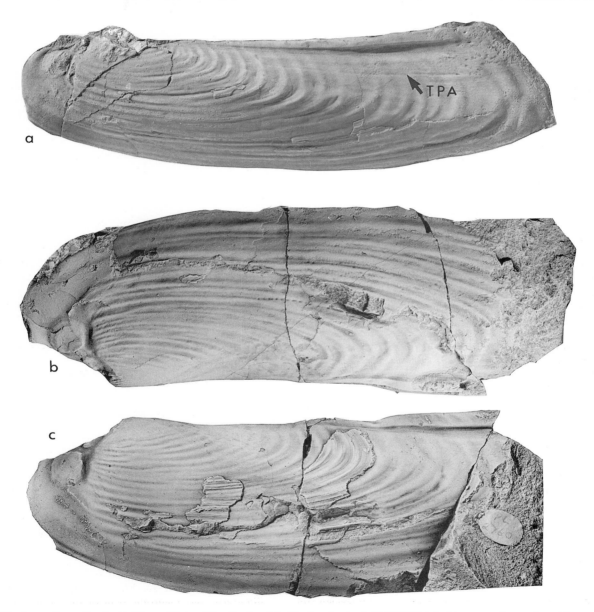

Fig. 45 *Siliquimya plicata* (Portlock). Upper Viséan. Fig. 45a, Lowick, Northumberland; SM E2817, left side, trace of posterior adductor clearly visible (TPA). Figs 45b–c, near Keswick, Cumbria; BM PL2760; Fig. 45b, right side of steinkern; Fig. 45c, left side. All ×1.

Asbian, of Redesdale, Northumberland. BM L8988 from the Viséan, Wardle Shale, near Edinburgh. BM L47507 from the Viséan, Upper Limestone, at Orchard, near Glasgow. BM L46479 from the Viséan, at Lawston Linn, Liddet Water. SM E2816–7, E20863 and E1046 from Lowick, Northumberland.

DESCRIPTION. An elongate, compressed, soleniform species with low umbones well towards the anterior. The anterior margins are subrounded and the posterior margins are obliquely truncated, sloping backwards, and form an angle with the dorsal margins, but are rounded ventrally. Some Redesdale Ironstone specimens have the shell preserved, apparently without periostracal spicules. The ornament consists of comarginal rugae which are offset and broader along a line between the umbo and the posterior ventral margin. The ligament is posterior, set on a long slender nymph. The dorsal margins are linear and contiguous to the posterior margin. The ligament is set down in a long, narrow carinate escutcheon; the carinae are sharp and very gently concave. The anterior adductors are subcircular and relatively small with a considerable thickening of the shell behind them in the form of a low, straight, slightly anterior sloping clavicle. The posterior adductors are hardly visible in most specimens; they lie well towards the posterior margin and close to the dorsum; they are subcircular. The pallial line is moderately incised at the anterior, but it is very difficult to follow towards the posterior. Hind (1900: 388) interpreted it, apparently correctly, as entire.

INCERTAE SEDIS

Genus *SPHENOTUS* Hall, 1885

TYPE SPECIES. *Sphenotus arcaeformis* (Hall & Whitfield, 1869), subsequently designated by Miller, 1889: 513, Middle Devonian, Hamilton, New York State. For a figure see Hall, 1885: pl. 65, figs 7–11. There are no modern illustrations of this species.

The generic name *Sphenotus* has been used by Driscoll (1965) and Pojeta (1969) for species we would include in *Pleurophorella*. Neither author based his opinion on a reconsideration of Hall & Whitfield's type species, which is of Middle Devonian age. Examination of Hall's illustrations suggests that these may be related genera but we do not find sufficient similarity to accept their synonymy. McAlester's Upper Devonian *Sphenotus tiogenesis* may be a thin-shelled mud-dwelling member of this genus (McAlester, 1962: 62; pl. 26, figs 1–14).

Genus **GRAMMYSIOIDEA** Williams & Breger, 1916
(See Runnegar, 1974: 931)

TYPE SPECIES. *G. princiana* Williams & Breger 1916: 133. from the Lower Devonian. Moose River. Miss. We have examined a syntype, USNM 66190, kindly lent by Mr F. Collier, which is badly crushed and distorted, and has no ornament preserved. This species was apparently quite wrongly illustrated in the *Treatise* (Newell 1969: N821). We conclude that at present the species and genus is unrecognizable and should be rejected as a *nomen dubium*.

A CLASSIFICATION OF THE ANOMALODESMATA

This study of Upper Palaeozoic Anomalodesmata has revealed a greater diversity than we previously suspected. It is the documentation of this diversity which we consider furnishes further information for the overall classification of the Anomalodesmata. Below we offer an interim classification of Upper Palaeozoic taxa which is a modification of Cox *et al.* (1969). Runnegar (1974) and Morton (1982). We expect to modify the classification further, when we study the Mesozoic taxa.

Subclass **ANOMALODESMATA** Dall, 1889

We interpret the characters of primitive members to include an aragonitic nacreoprismatic shell with periostracal spicules developed early in their history. The hinge consists of a slender hinge plate with an opisthodetic parivincular ligament set on slender nymphs and few or no hinge teeth. An escutcheon is usually present and the posterior dorsal margins are close and joined by periostracum. Adductors subequal to anisomyarian. Pallial line primitively without sinus. Shapes typical of sessile deep burrowers to byssate nestlers.

?Order **ORTHONOTOIDA** Pojeta, 1978

Superfamily **ORTHONOTACEA** Miller, 1877
[nom. trans. Pojeta, 1978]

Family **ORTHONOTIDAE** Miller, 1877

Elongate shells, apparently without periostracal spicules. Ligament external, opisthodetic.

Orthonota Conrad, 1841.
Palaeosolen Hall, 1885; gross shell characters convergent with the heterodont superfamily Solenacea.
?*Cymatonota* Ulrich, 1893.

Superfamily uncertain

Family **SOLENOMORPHIDAE** Paul, 1941
(=Solenopsidae Neumayr, 1883)

Elongate shells, with external opisthodetic ligament. Shell structure unknown. (The species *Solenomorpha elegantissima* Hayasaka, 1925 which has well preserved periostracal spicules has been better placed in *Alula* by Hayami & Kase 1977). An alternative classification would relate *Solenomorpha* to the elongate M. Devonian Sanguinolitinae with the lack of periostracal spicules interpreted as secondary loss. In that case Solenomorphidae would be closer to the mainstream anomalodesmatids.

Subfamily **SOLENOMORPHINAE** Paul, 1941
[nom. trans. herein]

Umbones towards or at the anterior. Deeper burrowing attained by elongation of the posterior shell.

Solenomorpha Cockerell, 1903; no spicules known.
Ennirostra Hajkr, Lukasova, Ruzicka & Rehor, 1975.

?Subfamily **PROMACRINAE** nov.

Radiating striae present, apparently without spicules. Elongation of shell anterior to umbones to give a *Donax*- or *Solemya*-like shape. Shell structure unknown, muscle scars poorly known.

Promacrus Meek, 1871.

?Family **PROTHYRIDAE** Miller, 1889

Dorsal part of anterior margins bear small protrusion.

Prothyris Meek (*in* Meek & Worthen 1869); fine radiating striae, no periostracal spicules.
Paraprothyris Clarke, 1913.
Amphikoilum Novozilov, 1956.

Order **PHOLADOMYOIDA** Newell, 1965
(?=Myoida Stoliczka, 1870;
=Desmodontida Neumayr, 1883)

Usually infaunal nestlers to very deep burrowers. Ligament primitively external, opisthodetic, becoming internal in several unrelated post-Palaeozoic lineages. Pallial line primitively without sinus, but developing this feature in several eparate lineages, the most advanced forms in this respect having long siphons of type 'C'. Shell surface bearing periostracal spicules in the primitive forms. Ordovician to Recent. A paraphylum including the ancestors of at least some septibranchs.

?Superfamily **EDMONDIACEA** King, 1850

Without surface spicules, buttressing parallel to hinge commonly present, characteristic pedal muscle scar pattern commonly present. Shell structure unknown. The Edmondiacea share shell shape and simplicity of hinge with the more primitive Pholadomyoida, characters which cannot be counted as firm synapomorphies. They do not possess the hypertrophied anterior adductor muscle, a synapomorphy of the Lucinoida, and it would be unreasonable to suggest that *Allorisma* had an anterior inhalent current which we would interpret as a primitive character possessed by most Lucinoida. Any arrangement of hinge teeth is more simple than that possessed by any of the major groups of Heteroconchia, so the Edmondiacea remain in the Pholadomyoida rather by default than by sharing any recognized synapomorphy.

Family **EDMONDIIDAE** King, 1850
(?=Cardiomorphidae Miller, 1877;
=Allorismidae Astafieva-Urbaitis, 1964)

Edmondia de Koninck, 1841.
Allorisma King, 1844.
Scaldia de Ryckholt, 1847.
Cardiomorpha de Koninck, 1841.

Family **MEGADESMIDAE** Vokes, 1967
(=Pachydomidae Fischer, 1886, nom. inval.)

Megadesmus J. de C. Sowerby, 1839.
Astartila Dana, 1847.
Pyramus Dana, 1847.
Plesiocyprinella Holdhaus, 1918.
Farrazia Cowper Reed, 1932.
?Casterella Mendes, 1952.

Superfamily **PHOLADOMYACEA** King, 1844
[nom. trans. Newell, 1965]
(=Grammysiacea Miller, 1877, nom. trans. Dickins, 1963)

Primitively myiform with external, posterior parivincular ligament mounted on paired upward-facing nymphs. Primitive shell structure considered to be nacreo-prismatic aragonite with radiating rows of periostracal spicules. Hinge line with few or no hinge teeth. Usually elongate shells with rounded or sub-rounded ends, often with a subumbonal sulcus. Shallow to deep sessile burrowers. Although a pallial sinus is present in many Upper Palaeozoic taxa, the more primitive lack this feature. We may interpret from this that they were primitively without siphons, but siphons of varying complexity, types 'B' or 'C', apparently develop separately in a number of lineages. The more primitive living forms are eulamellibranch filter feeders. A paraphylum including the Pholadomyidae and their Palaeozoic ancestors, together with the ancestors of the Thraciacea, Pandoracea, Poromyacea, Hiatellacea, Gastrochaenacea, Pholadidacea, Clavagellacea and probably the Myacea.

Family **GRAMMYSIIDAE** Miller, 1877

Sulcate forms with a break in shell ornament, becoming arcticiform. Ligament external, born on narrow nymphs. Shell structure unknown but surface commonly with radiating lines of periostracal spicules.

Subfamily **GRAMMYSIINAE** Miller, 1877
[nom. trans. herein]

Later taxa arcticiform, pallial line without sinus. Sulcus usually present.

Grammysia de Verneuil, 1847.

Subfamily **CUNEAMYINAE** nov.

Elongate, myiform, pallial line incompletely known. Shallow subumbonal sulcus sometimes present.

Cuneamya Hall & Whitfield, 1875.
?Rhytimya Ulrich, 1884.
?Grammysioidea Williams & Breger, 1916.
?Protomya Hall, 1885 (=*Palaeomya* Hall, *non* Zittel & Goubert, 1861).

Family **SINODORIDAE** Projeta & Zhang, 1984
[Elevated to a superfamily by Pojeta, Zhang & Yang, 1986.]

Sinodora Pojeta & Zhang, 1984.
Palaeodora Fleming, 1957.

Family **SANGUINOLITIDAE** Miller, 1877

Pallial sinus absent to deep. Ligament external opisthodetic, mounted on nymphs. Non-gaping to widely gaping. Hinge teeth usually absent but 'cardinals' known to be present in Alulinae. Nacreo-prismatic shell structure known in some sanguinolitines, probably occurred throughout the family.

Subfamily **SANGUINOLITINAE** Miller, 1877
(?=Arcomyidae Fischer, 1886)

Pallial sinus shallow or absent. Shell elongate. No hinge teeth.

Sanguinolites M'Coy, 1844.
Myofossa Waterhouse, 1969*b*.
Palaeocorbula Cowper Reed, 1932.
Ragozinia Muromzeva, 1984.
Cosmomya Holdhaus, 1913.
Grammysiopsis Chernychev, 1950.
Pentagrammysia Chernychev, 1950.
?Siphogrammysia Chernychev, 1950.
?Glossites Hall, 1885.
Cimitaria Hall & Whitfield, 1875.
Gilbertsonia gen. nov. (see p. 70).
?Pachymyonia Dun, 1932.
?Leinzia Mendes, 1949.

Subfamily **PHOLADELLINAE** Miller, 1887

Radial ribbing present.

Pholadella Hall & Whitfield, 1869.

Subfamily **ALULINAE** Mailleux, 1937

Median tooth present in RV only. Deeper burrowing attained by elongation of the posterior shell, convergent with Solenomorphidae.

Alula Girty, 1912.
Unklesbyella Hoare, Sturgeon & Kindt, 1979.
?Tellinomorpha de Koninck, 1885.

Subfamily **UNDULOMYINAE** Astafieva-Urbaitis, 1973

Deep pallial sinus known in some genera. Narrow anterior and or posterior gape sometimes present.

Wilkingia Wilson, 1959.
Praeundulomya Dickins, 1957.
?Manankovia Astafieva-Urbaitis, 1984.
Undulomya Fletcher, 1946.
Exochorhynchus Meek & Hayden, 1865.

Subfamily **CHAENOMYINAE** Waterhouse, 1966

Pallial line truncated by broad shallow sinus parallel to the vertical posterior margins; posterior gape wide and rounded.

Chaenomya Meek, 1865.

Subfamily **VACUNELLINAE** Astafieva-Urbaitis, 1973

Pallial line usually truncated with shallow to medium pallial sinus. Narrow posterior gape often present.

Vacunella Waterhouse, 1965.
?Australomya Runnegar, 1969.
Myonia Dana, 1847.

Family **PERMOPHORIDAE** van de Poel, 1959
[*nom. nov. pro* Pleurophoridae Dall, 1895]
(?=Kalenteriidae but not including Redoniidae Babin, 1966)

Elongate ovate of modioliform with external opisthodetic ligament usually mounted on narrow nymphs. Periostracal spicules present only in the more primitive forms. Usually not gaping. Cross-lamellar shell structure known in Jurassic taxa.

Subfamily **PERMOPHORINAE** van de Poel, 1959

Permophorus Chavan, 1954.
Pleurophorella Girty, 1904.
?Pleurophorina Licharew, 1925.
Siliquimya gen. nov. (see p. 90).
Bowlandia gen. nov. (see p. 88).
Ivanovia Astafieva-Urbaitis, 1978.
?Cosmogoniophorina Isberg, 1934.
?Cosmogoniophora McLearn, 1918.
?Goniophorina Isberg, 1934.
?Naiadopsis Mendes, 1952.
?Jacquesia Mendes, 1944.
?Macackia Mendes, 1954.
?Roxoa Mendes, 1952.

Other genera as listed in the *Treatise*, except for *Redonia* Rouault, 1851 which is unlikely to belong to the Anomalodesmata or the Heteroconchia.

Eager, 1978, discussed the evolutionary origins of the Anthracosiacae. His hypothesis included an ancestor for that superfamily among late Viséan, apparently marine taxa, which he called *Sanguinolites* Hind, *non* M'Coy. He specifically mentioned two taxa, *Sanguinolites abdenensis* and *Sanguinolites ovalis*. These species, described by Hind (1900), are not well preserved and details of their hinge and musculature are not fully known. One possibility is that they belong to *Pleurophorella* as interpreted in the present work (p. 84); in which case the Anthracosiacea could prove to be a non-marine offshoot of the Permophoridae.

NOMEN DUBIUM

Sphenotus Hall, 1885.

CONCLUSIONS

The Anomalodesmata were more prominent during the Upper Palaeozoic than in almost any modern environment, forming more than half of the total infaunal species in the British Viséan for example. However, they have slowly increased in numbers of species from the late Palaeozoic to the Recent. Their less prominent position today is purely the result of much more rapid diversification of other infaunal groups in the later Mesozoic and Tertiary, particularly the siphonate heterodonts.

Diagnosis of the subclass is difficult; it is recognized by particularly negative characters which include few or no hinge teeth and a generalized myiform shell. The parivincular ligament borne on nymphs, clearly primitive for the group, is shared by the Heteroconchia and the more primitive nuculoids. We recognize periostacal spicules as a primitive character for the mainstream Anomalodesmata that include the Pholadomyoida. This leaves us with considerable uncertainty as to which bivalves are the closest sister groups of this order.

The traditional inclusion of the Edmondiacea within the subclass and the disputed inclusion of the Orthonotida are neither confirmed nor denied by any evidence we have been able to find. The superficial resemblance between *Allorisma* (Edmondiacea) and the Undulomyinae is shown to be a case of convergence.

Our classification has made use of more taxa at the family and subfamily level than some recent classifications of this group, e.g. Newell, 1969 and Runnegar, 1974. Although the Upper Palaeozoic anomalodesmatids did not exploit the variety of internal hinges typical of the Mesozoic and Kainozoic, their diversity of shell shape and pallial sinus, both reflecting their life habits and our interpretation of their phyletic relationships necessitate this action.

The Upper Palaeozoic subfamilies within the Sanguinolitidae differ essentially from the non-siphonate Grammysiidae, particularly including *Grammysia* itself, in all developing deeper burrowing siphonate forms. We have been able to establish polarity of characters of the dorsal shell margins and hinge within the Sanguinolitidae. We have related this to the evolution of siphons. We have interpreted an elongate carinate escutcheon and no posterior gape as primitive and loss of carinate escutcheon and acquisition of a posterior gape as advanced. This polarity has guided us in our taxonomic evaluation.

One interesting aspect of the Runnegar schematic view of anomalodesmatid evolution (1974: text-fig. 3) is that it shows absolutely no interruption at the Permo-Triassic boundary except for the demise of the Australasian taxon Megadesmidae. At this time we are uncertain whether or not the Megadesmidae may themselves be ancestral to at least some septibranchs. In the evidence as it is known, we can also show no distinct indication of an extinction event at this time but feel the record close to the boundary, and particularly in the early Triassic, is so poor that at present no reasonable interpretation can be made. The one possibility of an extinction at the family level at this time is the Edmondiidae, but we do not know whether the Mesozoic family Mactromyidae is similar because of common descent or because of convergence.

We have established the broad similarity between the Permophoridae and the Sanguinolitidae which we interpret as reflecting a close phyletic relationship. As byssate nestlers, crevice dwellers and at least one apparent cavicolous taxon, they foreshadow some of the habits of their post-Palaeozoic descendants, which we believe may include the Gastrochaenacea and *Hiatella*.

The present apparent poverty or patchiness of the Devonian record of Anomalodesmata leads to a number of uncertanties; e.g. we are unable to show whether or not the multiplicity of development of siphonate forms which is apparent by the Lower Carboniferous (Viséan) is a result of an earlier Carboniferous radiation with some convergence of shape to early Palaeozoic taxa or whether the individual clades, subfamilies in this study, have a more ancient history. The earlier Palaeozoic Subfamily Cuneamyinae includes taxa with similar shape to the Upper Palaeozoic siphonate ones but we have been unable to discover the nature of the pallial line, and hence presence or absence of siphons, in this early group, nor whether there were repeated parallel radiations producing convergent forms.

This is of particular importance in considering the tracing of their history from the Palaeozoic through to the Mesozoic. The schematic evolutionary tree outlined by Runnegar (1974) and repeated by Morton (1987) is an over-simplification and is replaced here by a classification that is both rather more complicated and less certain in some details. However, the essential aspect of shallow burrowers giving rise to deeper burrowers, which may be interpreted from a comparison of shell morphology, remains a key insight into their evolution. Our own classification is outlined above.

The stratigraphical distribution within the Upper Palaeozoic of anomalodesmatids has been used for correlation, particularly in the early Permian of eastern Australia. We find further stratigraphical value in the Sanguinolitidae, particularly those with prominent or discordant ornament. We are, however, perplexed by, and unable to resolve without further field collecting, the bivalve fauna from the Upper Palaeozoic shale sequence from Peru, described by Thomas (1928). This has a distinct Upper Artinskian aspect but is accompanied in the collections at Cambridge and in Thomas' description by Pennsylvanian ammonoids. The distinctive part of the fauna includes species of *Ragozinia*, *Undulomya* and *Exochorhynchus*. Does this represent an earlier occurrence of taxa in an as yet unrecognized southern bivalve province, or is it the result of the mixing of two faunas, possibly when they were collected?

ACKNOWLEDGEMENTS. We would like to thank the following for access to collections and loan of specimens, or general discussion: B. M. Bell formerly at the New York State Museum, Albany; Norman Newell at the American Museum of National History; Murray Mitchell at the British Geological Survey; Ron Cleevely, Solene Morris, Phil Palmer and John Taylor and many other colleagues at the British Museum (Natural History); Annie Dhondt and Paul Sarteneer at the Institute d'Histoire Naturelle in Brussels; Stephanie Etchells Butler and Barry Rickards at the Sedgwick Museum, Cambridge; Colm o'Riorden formerly at the National Museum of Ireland, Dublin; Brian Morton of the University of Hong Kong; S. C. Shah at the Geological Survey of India, Chen Jing Hua and Fang Zongjie at the Institute of Geology and Palaeontology, Nanjing; Bruce Runnegar formerly at the Univerisity of New England; Jim Kennedy and Phil Powell at the University Museum, Oxford; Erle Kaufmann (formerly), and Fred Collier at the United States National Museum; Heinz Kohlmann at the Naturhistorisches Museum, Vienna; Barbara Pyrrah at the Yorkshire Museum. Hugh Owen in the Palaeontology Department of the BM(NH) kindly identified a number of Upper Palaeozoic ammonoids and we would like to thank Phil Crabb, Harry Taylor and others in the photo studio at the BM (NH) who took many of the photos.

REFERENCES

Agassiz, L. 1842–45. *Études critiques sur les Mollusques fossiles: Monographie des Myes.* xvii + 287 pp., pls 1a–39. Neuchâtel.

Aller, R. C. 1974. Prefabrication of shell ornamentation in the bivalve *Laternula*. *Lethaia*, Oslo, **7**: 43–56.

Astafieva-Urbaitis, K. A. 1962. [The genus *Allorismiella* gen. nov. in the Lower Carboniferous of the Sub-Moscow Basin.] *Izvestiya i Vўsshikh Uchebnўkh Zavedeniĭ. Geologiya i Razvedka*, Moscow, **1962** (2): 35–43 [In Russian].

—— 1964. [The genus *Allorisma* from the Lower Carboniferous of the Sub-Moscow Basin.] *Paleontologicheskiĭ Zhurnal*, Moscow, **1964** (1): 45–55 [In Russian].

—— 1970. Characteristics and systematic position of the bivalve genus *Edmondia*. *Paleontological Journal*, Washington **4** (3): 324–329 (Engl. transl. of *Paleontologicheskiĭ Zhurnal*, Moscow, **1970** (3): 41–47).

—— 1973. On the systematics of the Megadesmidae (Bivalvia). *Paleontological Journal*, Washington, **7** (1): 9–14 (Engl. transl. of *Paleontologicheskiĭ Zhurnal*, Moscow, **1973** (1): 13–19).

—— 1974a. Characteristics and systematic position of the bivalve genus *Sanguinolites* M'Coy. *Paleontological Journal*, Washington, **87** (1): 49–54 (Engl. transl. of *Paleontologicheskiĭ Zhurnal*, Moscow, **1974** (1): 54–60).

—— 1974b. The genus *Praeundulomya* in the Paleozoic of the USSR. *Paleontological Journal*, Washington, **8** (2): 152–156 (Engl. transl. of *Paleontologicheskiĭ Zhurnal*, Moscow, **1974** (2): 38–44).

—— 1978. *In* Astafieva-Urbaitis, K. A. & Ramovš, A., Verxnekamennougolvn'e (Gzhel'skie) Dvustvorki is 'Vornnchkogo Routa (Karavanki, Soloveni'). *Geologija. Rasprave in Poročila*, Ljubljana, **21** (1): 5–34, pls 1–3 [In Russian, English abstract].

—— 1981. The genus *Exochorhynchus* from the Upper Paleozoic deposits in the USSR and Mongolia. *Paleontological Journal*, Washington, **15** (3): 29–36 (Engl. transl. of *Paleontologicheskiĭ Zhurnal*, Moscow, **1981** (3): 35–42).

—— 1984. [*Manankovia* — a new genus of Carboniferous bivalves.] *Trudў Sovmestnaya Sovetska-Mongol'skaya Paleontologicheskaya Ekspeditsiya*, Moscow, **20**: 66–74. [In Russian].

—— & Dickins, J. M. 1984. Novў rod Kamennougol'mykh Sanguinolitid (Bivalvia) [A new genus of Carboniferous Sanguinolitids (Bivalvia)]. *Paleontologicheskiĭ Zhurnal*, Moscow, **1984** (3): 36–41. (Engl. transl. *Paleontological Journal*, Washington, **18** (3): 31–36 (1985)).

——, Lobanova, O. V. & Muromtseva, V. A. 1976. The genus *Myonia* (Bivalvia) in the Permian of the north-east of the USSR. *Paleontological Journal*, Washington, **10** (1): 23–36 (Engl. transl. of *Paleontologicheskiĭ Zhurnal*, Moscow, **1976** (1): 27–40).

Babin, C. 1966. *Mollusques bivalves et cephalopodes du Paleozoic Armoricain*. 471 pp. Brest (Imprimerie Commerc. Adm.).

Beede, J. W. 1902. Invertebrate paleontology of the red beds. *1st biennial report of the Oklahoma Geological Survey*, Norman, **Advance Bulletin**: 1–11.

Bernard, F. 1895. Première note sur le développement et la morphologie de la coquille chez les lamellibranches. *Bulletin de la Société géologique de la France*, Paris, **23**: 104–154.

Beushausen, L. 1895. Die Lamellibranchiaten des rheinischen Devon mit Ausschluss der Aviculiden. *Abhandlungen der Königlich Preussischen geologischen Landesanstalt*, Berlin, (N.F.) **17**. 514 pp., 38 pls.

Bittner, A. 1895. Lamellibranchiaten der Alpinen Trias. 1 Teil: Revision der Lamellibranchiaten von Sct. Cassian. *Abhandlungen der Kaiserlich-Königlichen geologischen Reichsanstalt*, Vienna, **18** (1): 1–135, pls 1–24.

Boyd, D. W. & Newell, N. D. 1968. Hinge grades in the evolution of Crassatellacean bivalves as revealed by Permian genera. *American Museum Novitates*, New York, **2328**. 52 pp.

Brown, T. 1837–49. *Illustrations of the Fossil Conchology of Great Britain and Ireland*. viii + 273 pp., 98 pls. London.

Carter, J. G. & Aller, R. C. 1975. Calcification in the bivalve periostracum. *Lethaia*, Oslo, **8**: 315–320.

Chavan, A. 1954. Les *Pleurophorus* et genres voisins. *Cahiers géologiques de Thoiry*, **22**: 200.

—— 1969. Superfamily Carditacea. *In* Moore, R. C. (ed.), *Treatise on Invertebrate Paleontology*, N (Mollusca 6, Bivalvia): N543–561. Lawrence, Kansas.

Chernychev, B. I. 1950. [The family Grammysiidae from the Upper Palaeozoic deposits of the U.S.S.R.] *Trudў Instituta Geologicheskikh Nauk. Seriya Stratigraphii i Paleontologii*, **1**. 91 pp., 14 pls. [In Russian].

Clarke, J. M. 1913. Fosseis Devonianos de Paraña. *Monografias Servico Geologico e Mineralogico Brasil*, Rio de Janeiro, **1**: 1–353, 27 pls.

Cockerell, T. D. A. 1903. Some homonymous generic names. *Nautilus*, Philadelphia, **16**: 116.

Conrad, T. A. 1841. Fifth annual report on the Palaeontology of the State of New York. *In: Fifth annual report of the Geological Survey, State of New York*: 25–57.

Cowper Reed, F. R. 1932. New fossils from the Agglomerate Slate of Kashmir. *Memoirs of the Geological Survey of India, Palaeontologia Indica*, Calcutta, **20** (1). ii + 79 pp., 13 pls.

Cox, L. R. 1969. *In* Moore, R. C. (ed.), *Treatise on Invertebrate Paleontology*, N (Mollusca 6, Bivalvia): N491–952. Lawrence, Kansas.

Dall, W. H. 1889. On the Hinge of Pelecypods and its development with an attempt toward a better subdivision of the Group. *American Journal of Science*, New Haven, (3) **38**: 445–462.

—— 1895. Contributions to the Tertiary fauna of Florida. *Transactions of the Wagner Free Institute of Science*, Philadelphia, **3** (3): 483–570.

Dana, J. D. 1847. Descriptions of Fossil Shells of the Collections of the Exploring Expedition under the command of Charles Wilkes, U.S.N., obtained in Australia. . . . [&c.] *American Journal of Science*, New Haven, (2) **4**: 151–160.

Dickins, J. M. 1956. Permian Pelecypods from the Carnarvon Basin, Western Australia. *Bulletin of the Bureau of Mineral Resources, Geology and Geophysics*, Canberra, **29**. 42 pp., 6 pls.

—— 1957. Lower Permian pelecypods and gastropods from the Carnarvon Basin, Western Australia. *Bulletin of the Bureau of Mineral Resources, Geology and Geophysics*, Canberra, **41**. 56 pp., 10 pls.

—— 1963. Permian pelecypods and gastropods from Western Australia. *Bulletin of the Bureau of Mineral Resources, Geology and Geophysics*, Canberra, **63**. 150 pp., 26 pls.

—— & Shah, S. C. 1965. The pelecypods *Undulomya*, *Cosmomya* and *Palaeocosmomya* in the Permian of India and Western Australia. *Journal of the Geological Society of Australia*, Adelaide, **12**: 253–260, 2 pls, 1 fig.

Driscoll, E. G. 1965. Dimyarian pelecypods of the Mississippian Marshal Sandstone of Michigan. *Palaeontographica Americana*, Ithaca, **5** (35). 128 pp., 5 pls.

Dun, W. S. 1932. The lower marine forms of *Myonia*, with notes on a proposed new genus, *Pachymyonia*. *Records of the Australian Museum*, Sydney, **18** (8): 411–414, pls 51–52.

Eager, M. 1978. Discussion. *In* Morris, N. J. 1978 (q.v.). *Philosophical Transactions of the Royal Society of London*, (B) **284**: 274–275.

Elias, M. K. 1957. Late Mississippian fauna from the Redoak Hollow Formation of Southern Oklahoma. Part 3, Pelecypoda. *Journal of Paleontology*, Tulsa: **31** (4): 737–784.

Etheridge, R. 1877. Further contributions to Carboniferous Palaeontology. *Geological Magazine*, London, (2) **4**: 214–251, 306–309, pls 12–13.

Fedotov, D. M. 1932. Kanennougolviye Plastinchatozhaberiye Molluski Donechkogo Bassenia. [The Carboniferous Pelecypods of the Donetz Basin.]. *Trudў Vsesoyuznogo Geologo-Rasvedochnogo Ob'edineniye N.K.T.P., S.S.S.R.*, Leningrad, **103**. vi + 241 pp., pls 1–18, A, B. [In Russian].

Fischer, P. 1880–87. *Manuel de Conchologie et de Paléontologie conchyliologique*. xxv + 1369 pp. Paris.

Fleming, C. A. 1957. Lower Devonian Pelecypods from Reefton, New Zealand. *Transactions of the Royal Society of New Zealand*, Dunedin, **85** (1): 135–140.

Fleming, J. 1828. *A History of British Animals . . . [&c.].* xxiii + 565 pp. Edinburgh & London.

Fletcher, H. O. 1946. New Lamellibranchia from the Upper Permian of Western Australia. *Records of the Australian Museum*, Sydney, **17** (2): 53–75, pls 25–29.

Geinitz, H. B. 1880. Nachträge zur Dyas, I. *Mittheilungen aus dem Königlichen Mineralogisch-Geologischen und Praehistorischen Museum in Dresden*, Cassel, **3**. 43 pp., 7 pls.

Girty, C. H. 1904. New Molluscan genera from the Carboniferous. *Proceedings of the United States National Museum*, Washington, **27** (1372): 721–736.

—— 1910. New genera and species of Carboniferous fossils from the Fayetteville Shale of Arkansas. *Annals of the New York Academy of Sciences*, **20** (3) (2): 189–238.

—— 1912. On some invertebrate fossils from the Lykins Formation of Eastern Colorado. *Annals of the New York Academy of Sciences*, **22**: 1–8, pl. 1.

Gray, J. E. 1847. A list of the genera of Recent Mollusca, their synonyms and types. *Proceedings of the Zoological Society of London*, **15**: 129–219.

Hajkr, O., Lukasová, A., Růžička, B. & Řehoř, F. 1975. Aussonderung der Karbonmuschel *Ennirostra* gen. nov. als neue Gattung. *Spisy pedagogické fakulty v Ostravě:* **33**. 108 pp., 7 pls.

Hall, J. 1852. Notes upon some of the fossils collected on the route from the Missouri River to the Great Salt Lake, and in the vicinity of the latter place, by the expedition under the command of Captain Howard Stansbury, T. E. *Stansbury's Exploration and Survey of the Valley of the Great Salt Lake, Utah*, **Appendix E**: 407–414, pls 1–4.

—— 1875. Fossils of the Hudson River Group (Cincinnati Formations). *Report of the Ohio Geological Survey*, **2** (2): 67–161, pls 1–13.

—— 1885. Lamellibranchiata, II. Description and figures of the Dimyaria of the Upper Helderberg, Hamilton, Portage and Chemung Groups. *New York Geological Survey, Paleontology*, **5**, (1): 269–561.

—— & Whitfield, R. P. 1869. *Preliminary notice of the Lamellibranchiate shells of the Upper Helderberg, Hamilton, and Chemung Groups with others from the Waverly Sandstone*. Part 2. 88 pp. Albany. [Anonymously published.]

Hayami, I. & Kase, T. 1977. A systematic survey of the Paleozoic and Mesozoic Gastropoda and Paleozoic Bivalvia from Japan. *Bulletin University Museum, University of Tokyo*, **13**. 154 pp., 11 pls.

Hayasaka, I. 1925. On some Paleozoic molluscs of Japan. I. Lamellibranchiata and Scaphopoda. *Science Reports of the Tôhoku Imperial University*, Sendai, Ser. 2, **8** (2). 29 pp. 2 pls.

Hind, W. 1896–1905. A Monograph of the British Carboniferous Lamellibranch-iata, Part I (1896–1900), 476 pp., 56 pls. Part II (1901–1905), 216 pp, 25 pls. *Monographs of the Palaeontographical Society*, London.

Hoare, R. D., Sturgeon, M. T. & Kindt, E. A. 1979. Pennsylvanian marine Bivalvia and Rostroconchia of Ohio. *Bulletin of the State of Ohio Department of Natural Resources, Division of Geological Survey*, Columbus, **67**. 77 pp., 18 pls.

Holdhaus, K. 1913. Fauna of the Spiti Shales (Lamellibranchiata and Gastropoda). *Memoirs of the Geological Survey of India, Palaeontologia Indica*, Calcutta, (15) **4** (4): 397–456, pls 94–100.

—— 1918. Sobre alguns lamellibranchios fosseis do sul do Brazil. *Monographias servico Geologico e Mineralogico do Brasil*, Rio de Janeiro, **2**: 1–24, pls 1–2.

Isberg, O. 1934. *Studien über Lamellibranchiaten des Leptaenakalkes in Dalarna*. 492 pp., 9 text-figures, 32 pls. Lund.

Keyserling, A. F. M. L. A. von 1846. *Wissenschaftliche Beobachtungen auf einer Reise in das Petschora-Land im Jahre 1843*. iii + 465 pp., 22 pls, 2 maps col. St Petersburg.

King, W. 1844. On a new genus of Palaeozoic shells. *Annals and Magazine of Natural History*, London, (1) **14** (92): 313–317.

—— 1850. A Monograph of the Permian fossils of England. xxxviii + 258 pp., 29 pls. *Monographs of the Palaeontographical Society*, London.

Kiparisova, L. D., Bychkov, Y. M. & Polubotko, I. V. 1966. *Pozdnetriasovyye Dvustvorchatyye Mollyuski Severo-Vostoka SSSR [Late Triassic bivalved molluscs of the N.E. U.S.S.R.]*. 229 pp. 35 pls. Vsesoyuznogo Nauchno-Issledovatel'skogo Geologicheskogo Instituta (VSEGEI), Magadan. [In Russian].

Koninck, L. G. de 1841–44. *Déscription des animaux fossiles qui se trouvent dans le terrain carbonifère de Belgique*. 650 pp., pls A-H, 1–55. Liège.

—— 1877–78. Recherches sur les fossiles paléozoïques de la Nouvelle-Galles du Sud (Australia). Parts 1 & 2. *Mémoires de la Société Royale des Sciences de Liège*, (2) **6** (2). 140 pp., 4 pls (1877). Part 3. *Loc. cit.* **7** (1). 235 pp., 24 pls (1878).

—— 1885. Faune du Calcaire Carbonifère de la Belgique; Cinquième partie, Lamellibranches. *Annales du Musée Royal d'Histoire Naturelle de Belgique*, Brussels, **11**. 283 pp., 41 pls.

Krotova, P. 1885. Artinskii yrus. Geologo-Paleontologicheska Monography Artinskago Peschanika [Geological and palaeontological monograph of the Artinskian Sandstone]. *Trudÿ Obshchestva Estestviospÿtatelei pri Imperatorskom Kazanskom Universitete*, Kazan, **13** (5): 1–314, 4 pls.

Lamarck, J. B. 1818. *Histoire naturelle des animaux sans vertèbres*, **5**: 612 pp.; **6** (1): 343 pp. Paris.

Licharew, B. 1925. Zur Frage über des Alter der Perm-Kalkstein der Onega–Dwina Wasserscheide. *Zapiski Rossiiskogo Mineralogicheskogo Obshchestva*, Petrograd, **54** (1): 109–152.

Logan, A. 1964. The dentition of the Durham Permian pelecypod *Permophorus costatus* (Brown). *Palaeontology*, London, **2**: 281–285, pl. 47.

—— 1967. The Permian Bivalvia of Northern England. 72 pp., 10 pls. *Monographs of the Palaeontographical Society*, London.

Lutkevich, E. M. & Lobanova, O. V. 1960. Pelecipody Permi Sovietskogo Sektora Arktiki [The Permian Bivalves of the Soviet part of the Arctic]. *Trudÿ Vsesoyuznogo Neftyanogo Nauchno-Issledovatel'skogo Geologo-Razvedochnogo Instituta (V.N.I.G.R.I.)*, Leningrad, **149**: 1–294, pls 1–35.

McAlester, A. L. 1962. Upper Devonian pelecypods of the New York Chemung Stage. *Bulletin of the Peabody Museum of Natural History*, New Haven, Conn., **16**. 88 pp., 32 pls.

—— 1966. Evolutionary and systematic implications of a transitional Ordovician Lucinoid bivalve. *Malacologia*, Ann Arbor, **3** (3): 433–439.

McChesney, J. H. 1860. *Description of new species of Fossils from the Palaeozoic Rocks of the Western States* [1859]. 76 pp., pls. Chicago.

—— 1867. Descriptions of Fossils from the Palaeozoic Rocks of the Western States, with Illustrations. *Transactions of the Chicago Academy of Sciences*, **1**: 1–57, 9 pls.

McClearn, F. H. 1918. The Silurian Arisaig Series of Arisaig, Nova Scotia. *American Journal of Science*, New Haven, Conn., (4) **45**: 126–140.

M'Coy, F. 1842. *In* Griffith, R., *Notice respecting fossils of the Mountain Limestone*. 411 pp. Dublin.

—— 1844. *A synopsis of the Carboniferous Limestone fossils of Ireland*. viii + 207 pp., 29 pls., text illust. Dublin, privately printed by S. R. J. Griffith.

—— 1851a. Descriptions of some new Mountain Limestone Fossils. *Annals and Magazine of Natural History*, London, (2) **7**: 167–175.

—— 1851b, 1852 & 1855. A systematic description of the British Palaeozoic Fossils in the Geological Museum of the University of Cambridge. *In* Sedgwick, A., *A synopsis of the classification of the British Palaeozoic rocks*. 1851: i–iv, 1–184; 1852: i–x, 185–406; 1855: i–xcviii, 407–662. London & Cambridge.

Maillieux, E. 1937. Les lamellibranches du Dévonien Inférieur de l'Ardenne.

Mémoires du Musée royal d'Histoire Naturelle de Belgique, Brussels, **81**. 274 pp., 14 pls.

Meek, F. B. 1865. *In* Meek, F. B. & Hayden, F. (q.v.).

—— 1871. Descriptions of some new types of Paleozoic Shells. *American Journal of Conchology*, Philadelphia, **7**: 4–10, 1 pl.

—— 1873. Descriptions of Invertebrates from Carboniferous System. *Paleontology of Illinois. Geological Survey of Illinois*, (3) **5** (2): 323–619, 32 pls.

—— & Hayden, F. 1859. Remarks on the Lower Cretaceous Beds of Kansas and Nebraska, together with descriptions of some new species of Carboniferous fossils from the valley of Kansas River. *Proceedings of the Academy of Natural Sciences*, Philadelphia, **1858**: 256–266.

—— 1865. Palaeontology of the Upper Missouri. Invertebrates. Part I. *Smithsonian Contributions to Knowledge*, Washington, **14** (172): 1–135, pls I–V.

—— & Worthen, F. 1869 ('1870'). Descriptions of new Carboniferous fossils from the Western States. *Proceedings of the Academy of Natural Sciences*, Philadelphia, **1870**: 137–172. [This paper was first published privately in 1869, according to Ruhof].

Mendes, J. C. 1944. Lamelibranquios Triassicos de Rio Claro (Estado de São Paulo). *Boletim da Faculdade de Filosofia. Ciencias e Letras, Universidade de São Paulo*, **45** (Geologia 1): 41–76, 2 pls.

—— 1949. Novos lamelibranquios fosseis de serie Passa Dois, sul do Brazil. *Boletim, Divisão de Geologia e Mineralogia, Brasil*, Rio de Janeiro, **133**. 40 pp., 5 pls.

—— 1952. A formacão Corumbatai na regiao di Rio Corumbatai (Estratigraphia e descricao dos lamelibranquios). *Boletim da Faculdade de Filosofia, Ciencias e Letras, Universidade de São Paulo*, **145** (Geologia 8). 119 pp., 4 pls.

—— 1954. Contribuicão a estratigrafia da serie Passa Dois no estada do Paraña. *Boletim da Faculdade de Filosofia, Ciencias e Letras, Universidade de São Paulo*, **175** (Geologia 10). 119 pp., 3 pls.

Miller, S. A. 1877. *The American Paleozoic Fossils. A Catalogue of the Genera and Species*. 253 pp. Cincinnati, Ohio, published by the author.

—— 1889. *North American geology and palaeontology for use of amateurs, students and scientists*. 664 pp. Cincinnati, Ohio.

Morris, J. 1854. *A Catalogue of British Fossils: comprising the Genera and Species hitherto described with references to their geological distribution and to the localities in which they have been found*. (2nd edn.) viii + 372 pp. London, privately published by the author.

—— & Lycett, J. 1853–55. A Monograph of the Mollusca from the Great Oolite, chiefly from Minchinhampton and the coast of Yorkshire. Part 2, Bivalves. 1853: 1–80, pls I–VIII; 1855: 81–147, pls IX–XV. *Monographs of the Palaeontographical Society*, London.

Morris, N. J. 1978. The infaunal descendants of the Cycloconchidae: an outline of the evolutionary history and taxonomy of the Heteroconchia, superfamilies Cycloconchacea to Chamacea. *Philosophical Transactions of the Royal Society of London*, (B) **284**: 259–275.

Morton, B. 1982. The functional morphology of *Parilimya fragilis* (Bivalvia, Parilimyidae *nov. fam.*) with a discussion on the origin and evolution of the carnivorous septibranchs and a reclassification of the Anomalodesmata. *Transactions of the Zoological Society of London*, **36** (3): 159–236, illus.

—— 1987. Siphon structure and prey capture as a guide to affinities in the abyssal septibranch Anomalodesmata. *Sarsia*, Bergen, **72**: 49–69.

Muromzeva, V. A. 1974. Dvustvorchatye Mollyuski Karbona Kazakhstana i Sibiri [Bivalve molluscs of the Carboniferous of Kazakhstan and Siberia]. *Trudÿ Vsesoyuznogo Neftyanogo Nauchno-Issledovatel'skogo Geologo-Razvedochnogo Instituta*, Leningrad, **336**: 1–150, 29 pls. [In Russian].

—— 1984. Permskie Morskie Otozheni' i Dvustvorchatye Mollyuski Sovetskoi Artiki. *Trudÿ Vsesoyusnyi Ordena Trudovogo Krasnogo Zhameni Neftyuoi Nauchno-Isseldovatelvskii Geolorazvedochnyi Institut*, **Unnumbered Trudÿ**. 154 pp., 53 pls. Moscow (Ministry of Geology for the U.S.S.R.) [In Russian].

Nechaev A. W. 1894. [Die Fauna der permischen Ablagerungen des östlischen Theils des europäischen Russlands.] *Trudÿ Obshchestva Estestvoispÿtatelei pri Imperatorskom Kazanskom Universitete*, Kazan, **27** (4). 503 + 12 pp., 12 pls. [In Russian; German and Russian plate captions].

Neumayr, M. 1883. Zur Morphologie des Bivalvenschlosses. *Sitzungsberichte der Akademie der Wissenschaften. Mathematisch-Naturwissenschaftlichen Classe*, Vienna, **88** (1): 385–419, 2 pls.

Newell, N. D. 1940. Invertebrate fauna of the Late Permian Whitehorse Sandstone. *Bulletin of the Geological Society of America*, New York, **51** (1): 261–336.

—— 1956. Primitive desmodont pelecypods of the Australian Permian. *American Museum Novitates*, New York, **1799**. 13 pp.

—— 1965. Classification of the Bivalvia. *American Museum Novitates*, New York, **2206**. 25 pp.

—— 1969. *In* Moore, R. C. (ed.), *Treatise on Invertebrate Paleontology*, N (Mollusca 6, Bivalvia 2): N491–952. Lawrence, Kansas.

—— & Boyd, D. W. 1975. Parallel evolution in early Trigoniacean Bivalves. *Bulletin of the American Museum of Natural History*, New York, **154** (2): 53–162.

Newton, R. B. 1906. Fossils from Singapore discovered by John B. Scrivenor, FGS, Geologist to the Federated Malay States. *Geological Magazine*, London, (5) **3**: 487–496, 1 pl.

Novozilov, V. 1956. Crustacées phyllopodes bivalves, 1. Leaidae. *Trudў Paleontologicheskogo Instituta, Akademiya Nauk SSR*, Moscow & Leningrad, **61**. 128 pp., 14 pls.

Orbigny, A. d' 1849. *Prodrome de Paléontologie Stratigraphique Universelle*, **1**. 394 pp. Paris.

Owen, G., Trueman, E. R. & Yonge, C. M. 1953. The ligament in Lamellibranchiata. *Nature*, London, **171**: 73.

Paul, H. 1941. Lamellibranchiata Infracarbonica. *Fossilium Catalogus*, Berlin, **91**. 348 pp.

Phillips, J. 1836. *Illustrations of the Geology of Yorkshire*, . . . [&c.], **Part 2**. xx + 253 pp., 24 pls., map. London.

—— 1848. The Malvern Hills compared with the Palaeozoic districts of Abberley, Woolhope, May Hill, Tortworth, and Usk. *Memoirs of the Geological Survey of Great Britain*, London, **2** (1): 386 pp. 30 pls.

Poel, L. van de 1959. Faune Malacologique du Hervien. *Bulletin de l'Institut Royal des Sciences Naturelles de Belgique*, Brussels, **35** (15): 1–26; (16): 1–28, 1 pl.

Pojeta, J. jr 1969. Revision of some of Girty's invertebrate fossils from the Fayetteville Shale (Mississippian) or Arkansas and Oklahoma – Pelecypods. *Professional Paper, United States Geological Survey*, Washington, **606C**: 15–24, pls 3–4.

—— 1971. Review of Ordovician pelecypods. *Professional Paper, United States Geological Survey*, Washington, **695**. 46 pp.

—— 1978. The origin and early taxonomic diversity of the pelecypods. *Philosophical Transactions of the Royal Society of London*, (B) **284**: 225–246.

—— & Zhang Renjie 1984. *Sindora* n. gen. – A Chinese Devonian homeomorph of Cenozoic Pandoracean Pelecypods. *Journal of Paleontology*, Tulsa, **58** (4): 1010–1025, 9 figs.

—— —— & Yang Zunyi 1986. Systematic Paleontology of the Devonian Pelecypods of Guangxi and Michigan. *Professional Paper, United States Geological Survey*, Washington, **1394G**: 57–108, pls 1–64.

Portlock, J. E. 1843. *Report on the Geology of the County of Londonderry and of parts of Tyrone and Fermanagh*. xxxii + 784 pp., 37 + 9 pls, map. Dublin and London (Geological Survey of Ireland).

Reed, see Cowper Reed.

Röding, P. F. 1798. *Museum Boltenianum* . . . (&c.) **2**. viii + 119 pp. Hamburg.

Rouault, M. 1850–51. Mémoire sur le terrain paléozoique des environs de Rennes. *Bulletin de la Société Géologique de France*, Paris, (2) **8**: 358–399.

Runnegar, B. 1965. The bivalves *Megadesmus* Sowerby and *Astartila* Dana from the Permian of eastern Australia. *Journal of the Geological Society of Australia*, Adelaide, **12** (2): 227–252, pls 12–15.

—— 1966. Systematics and biology of some desmodont bivalves from the Australian Permian. *Journal of the Geological Society of Australia*, Adelaide, **13** (2): 373–386.

—— 1967. Desmodont bivalves from the Permian of Eastern Australia. *Bulletin, Bureau of Mineral Resources, Geology and Geophysics, Australia*, Melbourne, **96**. 83 pp., 13 pls.

—— 1968. Preserved ligaments in Australian Permian bivalves. *Palaeontology*, London, **11** (1): 94–103.

—— 1969. Permian fossils from the southern extremity of the Sydney Basin. *In* Campbell, K. S. W. (ed.), *Stratigraphy and Palaeontology: Essays in Honour of Dorothy Hill*: 276–298. Canberra.

—— 1974. Evolutionary history of the bivalve class Anomalodesmata. *Journal of Paleontology*, Tulsa, **48** (5): 904–939.

—— & Newell, N. D. 1971. Caspian-like relict molluscan fauna in the South American Permian. *Bulletin of the American Museum of Natural History*, New York, **146** (1). 62 pp.

—— —— 1974. *Edmondia* and the Edmondiacea: shallow-burrowing Paleozoic pelecypods. *American Museum Novitates*, New York, **2533**. 19 pp.

Ryckholt, P. de 1851. Mélanges Paléontologiques. Part 1. *Mémoires Couronnés et Mémoires des Savants Etrangers. Académie Royale des Sciences, des Lettres et des Beaux-Arts de Belgique*, Brussels, **24** (2). 176 pp., 10 pls.

Sowerby, J. de C. 1827. *In*: Mineral Conchology of Great Britain. 7 vols, 648 pls. London.

—— 1839. Fossils. *In* Mitchell, T. L. (ed.), *Three expeditions into the interior of eastern Australia*. Revised edition, **2**. London.

—— 1840. *In* Prestwich, J. (ed.), On the geology of Coalbrookdale. *Transactions of the Geological Society of London*, (2) **5**: 413–495.

Stoliczka, F. 1870–71. Cretaceous fauna of southern India. *Memoirs of the Geological Survey of India, Palaeontologia Indica*, Calcutta, (6) **3**. 537 pp., 50 pls.

Taylor, J. D. 1973. The structural evolution of the bivalve shell. *Palaeontology*, London, **16** (3): 519–534.

——, Kennedy W. J. & Hall, A. 1973. The shell structure and mineralogy of the Bivalvia, II. Lucinacea–Clavagellacea, Conclusions. *Bulletin of the British Museum of Natural History*, London, (Zoology) **22**: 253–294.

Thomas, H. D. 1928. An Upper Carboniferous Fauna from the Amatope Mountains, North-Western Peru. *Geological Magazine*, London, **65**: 215–234, pls 6–8.

Ulrich, E. O. 1893. New and little known Lamellibranchiata from the Lower Silurian Rocks of Ohio and adjacent States. *Report of the Geological Survey of Ohio*, Columbus, **7**: 627–700, pls 45–56.

Verneuil, P. E. P. de 1845. *In* Murchison, R. I., Verneuil, P. E. P. de & Keyserling, A. (eds), *Geologie de la Russie d'Europe et des Montagnes d'Oural*, **2**. xxxii + 512 pp., 50 pls. London & Paris.

—— 1847. Note sur le parallélisme des roches dépôts paléozoïques de l'Amérique Septentrionale aves ceux de l'Europe, suivie d'un tableau des espèces fossiles communes aux deux continents, avec l'indication des étages où elles se rencontrent, et terminée par un examen critique de chacune de ces espèces. *Bulletin de la Société Géologique de la France*, Paris, (2) **4** (1): 646–710, table.

Vokes, H. E. 1967. Genera of the Bivalvia: a systematic and bibliographic catalogue. *Bulletin of American Paleontology*, Ithaca, **51** (232): 105–394.

Waterhouse, J. B. 1965. Generic diagnoses for some burrowing bivalves of the Australian Permian. *Malacologia*, Ann Arbor, **3** (3): 367–380.

—— 1966. On the validity of the Permian bivalve family Pachydomidae Fischer, 1887. *Journal of the Geological Society of Australia*, Adelaide, **13** (2): 534–559.

—— 1969a. The relationship between the living genus *Pholadomya* Sowerby and Upper Palaeozoic pelecypods. *Lethaia*, Oslo, **2**: 99–119.

—— 1969b. The Permian bivalve genera *Myonia*, *Megadesmus*, *Vacunella* and their allies, and their occurrence in New Zealand. *Palaeontological Bulletin, New Zealand Geological Survey*, Wellington, **41**: 1–141, 23 pls.

Weller, S. 1898. A bibliographic index of North American Carboniferous Invertebrates. *Bulletin of the United States Geological Survey*, Washington, **153**. 653 pp.

—— 1899. Kinderhook faunal studies 1. The Fauna of the Vermicular Sandstone at Northview, Webster County, Missouri. *Transactions of the Academy of Science of St Louis*, **9** (2): 9–31, pls 2–6.

Whidbourne, G. F. 1897. A Monograph of the Devonian Fauna of the South of England, 2. The Fauna of the Limestones of Lummaton, Woolborough, Chercombe Bridge and Chudleigh. 222 pp., 24 pls. *Monographs of the Palaeontographical Society*, London.

Williams, H. S. & Breger, C. L. 1916. The fauna of the Chapman Sandstone of Maine. *Professional Paper, United States Geological Survey*, Washington, **89**. 347 pp.

Wilson, R. B. 1959. *Wilkingia* gen. nov. to replace *Allorisma* for a genus of Upper Palaeozoic lamellibranchs. *Palaeontology*, London, **1** (4): 401–404, pl. 71.

—— 1960. A revision of the types of the Carboniferous lamellibranch species erected by J. Fleming. *Bulletin of the Geological Survey of Great Britain*, London, **16**: 110–124.

Yonge, C. M. 1957. Mantle fusion in the Lamellibranchia. *Pubblicazioni della Stazione Zoologica di Napoli*, **29**: 151–171.

Zittel, K. A. von & Goubert, E. 1861. Description des fossiles du Coral-Rag de Glos. *Journal de Conchyliologie*, Paris, **9**: 192–208.

SYSTEMATIC INDEX